# Elementary Quantum Mechanics

## With Problems and Solutions

# Elementary Quantum Mechanics

## With Problems and Solutions

Stephen Wiggins

*University of Bristol, UK*

**World Scientific**

NEW JERSEY · LONDON · SINGAPORE · BEIJING · SHANGHAI · HONG KONG · TAIPEI · CHENNAI · TOKYO

*Published by*

World Scientific Publishing Co. Pte. Ltd.

5 Toh Tuck Link, Singapore 596224

*USA office:* 27 Warren Street, Suite 401-402, Hackensack, NJ 07601

*UK office:* 57 Shelton Street, Covent Garden, London WC2H 9HE

Library of Congress Control Number: 2024948965

**British Library Cataloguing-in-Publication Data**
A catalogue record for this book is available from the British Library.

**ELEMENTARY QUANTUM MECHANICS**
**with Problems and Solutions**



ISBN 978-981-98-0185-5 (hardcover)
ISBN 978-981-98-0186-2 (ebook for institutions)
ISBN 978-981-98-0187-9 (ebook for individuals)

For any available supplementary material, please visit
https://www.worldscientific.com/worldscibooks/10.1142/14071#t=suppl

Typeset by Stallion Press
Email: enquiries@stallionpress.com

# Preface

This book has grown out of a one term course on quantum mechanics in the School of Mathematics at the University of Bristol. I have taught this course several times, but this book benefits immensely from the previous notes of Martin Sieber, Noah Linden, and Toby Cubitt. I am grateful that they have shared their materials with me. There is an extensive set of problems and solutions for this book. Many of the problems are my own, and some I have used in exams for the course. I have undoubtedly been influenced in my choice of problems by my own past courses in quantum mechanics, and some of those sources are mentioned in the book,

The book is intended for mathematics undergraduates. Generally, this means that the students do not have the background in physics to appreciate the issues surrounding the origin and development of quantum mechanics. One point that I try to emphasize is how new mathematics had to be developed in parallel with the quantum physics. In fact, "quantum mechanical thinking" has had a tremendous influence on mathematical thinking. A recent example of this is the fascinating review paper on linear stability analysis in fluid mechanics.[1] I hope that this tempts the students to study much of the physical background on their own with the supplemental references at the appropriate level that I give. Concerning prerequisites, I have tried to make the book as self-contained as possible. The amount

---

[1] Luca Magri, Peter J Schmid, and Jonas P Moeck. Linear flow analysis inspired by mathematical methods from quantum mechanics. *Annual Review of Fluid Mechanics*, 55(1):541–574, 2023.

of linear algebra required is not much beyond the level of computing the eigenvalues and eigenvectors of $2 \times 2$ matrices (with complex entries). The Schrödinger equation is a linear partial differential equation but I only consider potential energy functions that are independent of time. Therefore it can be analyzed using separation of variables. This is done *once*, and the necessary background can be developed in the context of this calculation. Consequently, only simple linear ordinary differential equations are required, which almost all students have seen before. I have taught this course in both the UK and the US to second, third, and fourth year mathematics students and prerequisites have not been an issue. In my experience, the material in this book can be covered in twelve weeks.

While this book originated during my time in the city of Bristol, United Kingdom, I have completed it in the past few months during my stay at the United States Naval Academy in Annapolis, Maryland. During this time I have used the book in teaching the course SM421, Elementary Quantum Mechanics, which was attended by third and fourth year students in mathematics. That class of midshipmen, Angela Feng, Nico Gonzalez-Reed, Jack Judy, Gabe Kitsch, James McAtee, Ivette Ayala, Cohen Bruner, Luke Gasper, Maya Kessner, Emily Rose Kosanovich, George Main, Will Nguyen, and Teagan Flinn, were particularly inquisitive and helpful in asking questions and I am grateful for all of their input and lively class participation. Spring of 2024 was a particularly challenging time for me and I am very grateful to Prof Kevin McIlhany of the department of physics at the Naval Academy for covering my lectures in the latter half of the course.

I am very grateful for the support of the William R. Davis '68 Chair in the Department of Mathematics at the United States Naval Academy which has helped make my stay at the academy very enjoyable and productive.

### *Books and Other Reference Materials*

There are literally hundreds of books on quantum mechanics, but the goals of this book are a bit different than the typical quantum mechanics textbook. While the origin of quantum mechanics is in physics, this book is intended for undergraduate mathematics majors whose background in physics may be quite limited. The intention of this book was to focus mostly on the mathematical structure and formalism of the subject, but to note the physical context in which the mathematics was developed without going into great detail. Numerous references for further reading are given throughout the book.

In terms of general references, below are some of my favorite that I have used over the years.

An excellent book on quantum mechanics for mathematics students is Hannabuss.[2] This book has some very nice choices of topics,

By the time of the late 1970's it was probably regarded by many physicists and chemists that quantum mechanics was a "mature" subject whose foundations were unlikely to change substantially in the future. This turned out to be far from true. Nevertheless, a book (actually, two volumes) first appeared in 1977[3] that was encyclopedic in its scope (for the time) and still remains today as an extremely useful text with excellent discussions about many topics: The sections on the mathematical foundations of quantum mechanics (including Dirac's notation), the postulates of quantum mechanics, angular momentum, and tensor products are all excellent (and all in the first volume). If you are going on to do further work in quantum mechanics then you will need to have a good reference on your shelf. This book is an excellent choice for that purpose. Many of the explanations are at a deeper level than you will find in most books.

David Bohm was one of the great thinkers about what quantum mechanics means . He wrote a book in the 1950's that was reprinted by Dover in 1989.[4] The sections on measurement and the Stern=Gerlach experiment are excellent.

I mentioned that by the late 1970s many viewed quantum mechanics as a "mature subject", but big changes were coming. Working on the foundations of quantum mechanics was not really a "mainstream" topic in the 1950's, 60's, and 70's, but this did not stop a number of pioneers. I mentioned Bohm, but Bell, Leggett, Clauser, Zurek are some other names that contributed fundamental ideas that are now definitely part of the "mainstream". The ability to test many of these seemingly abstract ideas exploded with experimental advances throughout the 1980's and 90's (continuing to this day, I touch on this in Chapter 5) and this has completely reinvigorated our thinking about the entire "structure" of quantum mechanics, and has likely also completely changed the way that it will be taught in the future. Textbooks are beginning to appear with this new

---

[2]K. Hannabuss. *An introduction to quantum theory*, volume 1. Clarendon Press, 1997.

[3]C. Cohen-Tannoudji, B. Diu, and F. Laloe. *Quantum Mechanics*. Wiley-VCH, 1992.

[4]D. Bohm. *Quantum theory*. Courier Corporation, 1951.

point of view, and an excellent example is the book by Schumacher and Westmoreland.[5]

Also notable is the book of David Tannor.[6] There are extremely useful topics that you will find in this book that you will not find in any other book, and they are developed in a very clear manner. One of the criticisms of beginning quantum mechanics courses is that students tend not to know how to "do anything" with it once they have finished the course. There is more than a little truth in this assessment. Tannor is a theoretical chemist, and theoretical chemists' job is to understand how quantum mechanical systems "do things". Tannor's book will show you how to use quantum mechanics to "get answers" about atomic and molecular systems.

### Internet Resources

The internet offers a wealth of (free) resources related to quantum mechanics. There is some very good material on the internet, and there is some material that is "not so good", so you should take care.

However, there is one internet source in quantum mechanics that I have used that is absolutely outstanding. These are the lecture notes of Professor Robert Littlejohn of the Physics department at the University of California at Berkeley. Professor Littlejohn himself has made fundamental contributions to quantum physics and he has developed these notes over the years in his course at Berkeley. They can be obtained here:

http://bohr.physics.berkeley.edu/classes/221/1011/221.html

Please reference them if you use them in any source. I will refer to them throughout these notes when I use them as "Littlejohn's notes". In general, these notes are at a more advance level than this course. However, I would particularly recommend Notes 1: The Mathematical Formalism of Quantum Mechanics, Notes 2: The Postulates of Quantum Mechanics, and Notes 13: Representations of the Angular Momentum Operators and Rotations. You will not find better discussions of angular momentum and spin, and their mathematical and physical descriptions, than exist in these notes.

---

[5]B. Schumacher and M. Westmoreland. *Quantum processes systems, and information.* Cambridge University Press, 2010.

[6]D. J Tannor. *Introduction to Quantum Mechanics: A Time-dependent Perspective.* University Science Books, 2007.

# Contents

# List of Figures

# Chapter 1

# The Mathematical Structure of Quantum Mechanics

In this chapter we develop the mathematics that is necessary to describe and compute the quantum mechanical concepts and quantities used in this book. In many cases we will not give complete proofs of mathematical results (especially in the infinite dimensional setting). The focus here will be describing the mathematical setting and results, and using them to describe and compute quantities in quantum mechanics.

## 1.1  Vector Spaces and Inner Products

We will learn that "quantum states" are points in an appropriately defined vector space. Therefore we begin by giving the definition of a vector space. This definition describes the rules for adding quantum states and multiplying quantum states by scalars (and in quantum mechanics the scalars are, in general, complex numbers).

**Definition 1 (Vector Space over a Field).** A vector space, denoted $V$, over a field $\mathcal{F}$, is a set of elements (the elements are called "vectors") on which two (binary) operations are defined:

Vector addition. For any $\phi$, $\psi \in V$, then $\phi + \psi \in V$.

Scalar multiplication. For any $\alpha \in \mathcal{F}$, $\phi \in V$, then $\alpha\phi \in V$.

Moreover, the operations of vector addition and scalar multiplication satisfy the following eight axioms.

Associativity of vector addition. $\psi + (\phi + \chi) = (\psi + \phi) + \chi, \quad \forall \psi, \phi, \chi \in V$.

Commutativity of vector addition. $\psi + \phi = \phi + \psi, \quad \forall \psi, \phi \in V$.

Identity element for vector addition ("existence of a zero vector"). There exists a vector $0 \in V$ such that $\psi + 0 = \psi, \quad \forall \psi \in V$.

Inverse element for vector addition. $\forall \psi \in V$ there exists a vector $-\psi \in V$ such that $\psi + (-\psi) = 0$.

Distributivity of scalar multiplication with respect to vector addition. $\alpha(\psi + \phi) = \alpha\psi + \alpha\phi, \quad \forall \alpha \in \mathcal{F}, \psi, \phi \in V$.

Distributivity of scalar multiplication with respect to field addition. $(\alpha + \beta)\psi = \alpha\psi + \beta\psi, \quad \forall \alpha, \beta \in \mathcal{F}, \psi \in V$.

Compatibility of scalar multiplication with field multiplication. $\alpha(\beta\psi) = (\alpha\beta)\psi, \forall \alpha, \beta \in \mathcal{F}, \psi, \in V$.

Identity element of scalar multiplication. There exists an element $1 \in \mathcal{F}$ such that $1\psi = \psi, \quad \forall \psi \in V$.

For the purposes of this book, the field $\mathcal{F}$ (i.e. scalars) will be the complex numbers. A vector space where the field is the complex numbers is often referred to as a *complex vector space*. The linear structure of quantum mechanics is a complex vector space.

We will also require an additional structure defined on a complex vector space–a complex inner product, which we now define.

**Definition 2 (Inner Product).** An inner product on a vector space $V$ is a map of an ordered pair of vectors, $(\psi, \phi)$ to the complex numbers that satisfies the following properties:

(i) $(\psi, \alpha\phi + \beta\chi) = \alpha(\psi, \phi) + \beta(\psi, \chi), \quad \alpha, \beta \in \mathbb{C}$.

(ii) $(\psi, \phi) = \overline{(\phi, \psi)}$.

(iii) $(\psi, \psi) \geq 0$, and equality holds if and only if $\psi = 0$,

for any $\alpha, \beta \in \mathbb{C}, \psi, \phi, \chi \in V$.

Using the inner product, we can construct a norm defined for vectors in $V$. Geometrically, the norm provides us with a measure of the "length" of vectors in $V$. The definition of the "norm induced by the inner product" is as follows.

**Definition 3 (Norm Induced by the Inner Product).** For any vector $\psi \in V$, we define the norm of $\psi$, denoted $\|\psi\|$, as follows:

$$\|\psi\| = \sqrt{(\psi, \psi)}. \tag{1.1}$$

Mathematically, norms satisfy certain properties (similar to inner products). In this case, those properties are inherited from the properties of the inner product. Rather than discussing in detail the properties satisfied by the norm, we will use Definition 3 without further comment.

We make the following remarks.

- Almost always, we will consider vectors $\psi \in V$ having norm one, i.e. $\|\psi\| = 1$. For any (nonzero) vector $\psi \in V$, we can divide it by its norm, and the result will be a vector having norm one. Such vectors are said to be *normalised*.[1]
- The vectors are said to be orthogonal if their inner product is zero, i.e. $\psi, \phi \in V$ are said to be orthogonal if $(\psi, \phi) = 0$. If $\psi$ and $\phi$ have norm one (or "unit norm") then they are said to be *orthonormal*.
- The complex inner product is *antilinear* (or "conjugate linear") in the first argument. The meaning of this phrase is made clear in the following calculation.

$$(\alpha\phi + \beta\chi, \psi) = \overline{(\psi, \alpha\phi + \beta\chi)} = \overline{\alpha\,(\psi, \phi) + \beta\,(\psi, \chi)}$$
$$= \bar{\alpha}\,(\phi, \psi) + \bar{\beta}\,(\chi, \psi), \tag{1.2}$$

which holds for any $\alpha, \beta \in \mathbb{C}$, $\psi, \phi, \chi \in V$.

- The Schwarz inequality (or Cauchy-Schwarz inequality, or Cauchy-Bunyakovsky-Schwarz inequality) is a fundamental inequality that has many uses in quantum mechanics. It bounds the magnitude of the inner product of two vectors by their norm, i.e.

$$|(\phi, \psi)| \leq \|\phi\|\|\psi\|, \quad \forall\psi, \phi \in V. \tag{1.3}$$

We prove this inequality. Consider the quantity:

$$(\psi + \alpha\phi, \psi + \alpha\phi) \geq 0. \tag{1.4}$$

Using the properties of the inner product to expand this expression gives:

$$(\psi + \alpha\phi, \psi + \alpha\phi) = (\psi + \alpha\phi, \psi) + \alpha(\psi + \alpha\phi, \phi),$$
$$= (\psi, \psi) + \bar{\alpha}(\phi, \psi) + \alpha(\psi, \phi) + \alpha\bar{\alpha}(\phi, \phi) \geq 0. \tag{1.5}$$

---

[1] "Normalizaton" will play a central role in our interpretation of the meaning of quantum mechanics in terms of observable quantities. We will see this in the next chapter.

Now we make the choice:

$$\alpha = -\frac{(\phi, \psi)}{(\phi, \phi)}. \tag{1.6}$$

Substituting (1.6) into (1.5) gives:

$$(\psi, \psi) - \frac{(\psi, \phi)(\phi, \psi)}{(\phi, \phi)} - \frac{(\phi, \psi)(\psi, \phi)}{(\phi, \phi)} + \frac{(\phi, \psi)(\psi, \phi)}{(\phi, \phi)} \geq 0. \tag{1.7}$$

or

$$(\psi, \psi) - \frac{(\psi, \phi)(\phi, \psi)}{(\phi, \phi)} \geq 0. \tag{1.8}$$

Multiplying (1.8) by $(\phi, \phi)$ gives:

$$(\psi, \psi)(\phi, \phi) \geq (\phi, \psi)(\psi, \phi), \tag{1.9}$$

which is the same as:

$$(\|\psi\| \, \|\phi\|)^2 \geq |(\phi, \psi)|^2, \tag{1.10}$$

which verifies the inequality.
- The norm satisfies the triangle inequality:

$$\|\psi + \phi\| \leq \|\psi\| + \|\phi\|, \quad \forall \psi, \, \phi \in V. \tag{1.11}$$

(We leave it as an exercise to show that this is true.)

Some examples of complex inner product spaces are the following.

*Example.* $\mathbb{C}^n$ denotes the space of $n$-tuples of complex numbers, $x \equiv (x_1, \ldots, x_n)$, $x_i \in \mathbb{C}$. It is straightforward to verify that $\mathbb{C}^n$ is a complex vector space. We define the following inner product on $\mathbb{C}^n$:

$$(x, y) = \sum_{j=1}^{n} \bar{x}_j y_j, \quad x, y \in \mathbb{C}^n. \tag{1.12}$$

We leave it as an exercise to verify that (1.12) satisfies the properties of an inner product.

*Example.* Let $L^2(D)$ denote the set of all complex valued functions of a real variable, $x$, defined on some domain $D \in \mathbb{R}$, denoted by $\psi(x)$, that satisfy the following:

$$\int_D |\psi(x)|^2 dx < \infty. \tag{1.13}$$

$L^2(D)$ becomes a complex vector space by defining vector addition and scalar multiplication "pointwise" as follows:

$$(\psi + \phi)(x) = \psi(x) + \phi(x), \qquad (1.14)$$

$$(\alpha\psi)(x) = \alpha\psi(x). \qquad (1.15)$$

We leave it as an exercise to verify that vector addition and scalar multiplication defined in this way satisfy the axioms of a complex vector space.

The complex vector space $L^2(D)$ can be made into a complex inner product space by defining an inner product on $L^2(D)$ as follows:

$$(\psi, \phi) = \int_D \overline{\psi(x)}\phi(x)dx. \qquad (1.16)$$

We leave it as an exercise to verify that (1.16) satisfies the properties of an inner product.

Thus far, our discussion of complex inner product spaces has been extremely general. In order to carry out the type of computations required in quantum mechanics, we need to introduce some additional structure on $V$. We begin with the notion of *linear independence*.

**Definition 4 (Linear Independence).** A set of vectors $\psi_1, \ldots, \psi_n \in V$ is said to be linearly independent if the only scalars satisfying the equation

$$\alpha_1\psi_1 + \cdots + \alpha_n\psi_n = 0,$$

are $\alpha_1 = \alpha_2 = \cdots = \alpha_n = 0$. Otherwise, the set of vectors is said to be linearly dependent.

The notion of linear independence leads naturally to the notion of *dimensionality* of a vector space.

**Definition 5 (Dimensionality).** A vector space is $n$-dimensional if it contains $n$ linearly independent vectors, but not $n + 1$. If it contains $n$ linearly independent vectors for every positive integer $n$ then it is infinite dimensional.

Next, we have the notion of a *spanning set* and a *basis* for a subset of $V$.

**Definition 6 (Spanning Set, Basis).** A set of vectors $\{\psi_1, \ldots, \psi_n\} \in V$ spans a finite dimensional vector space $W \subset V$ if every $\psi \in W$ can be written as $\psi = \alpha_1\psi_1 + \cdots + \alpha_n\psi_n$. A set of vectors $\{\psi_1, \ldots, \psi_n\} \in W$ is a basis for $W$ if they are linearly independent and they span $W$.

For infinite dimensional complex vector spaces we will need to give meaning to expressions like:

$$\sum_{i=1}^{\infty} \alpha_i \psi_i.$$

This notion is dealt with in the following definition.

**Definition 7 (Convergence).** A sequence of vectors $\psi_n \in V$ converges (strongly) to $\psi \in V$ if

$$\|\psi - \psi_n\| \to 0 \quad \text{as} \quad n \to \infty.$$

*A Typical Example in Quantum Mechanics of an Infinite Dimensional Complex Vector Space.* We consider the $2\pi$-periodic complex valued functions of a real variable, $\theta$, satisfying:

$$\int_0^{2\pi} |\psi(\theta)|^2 d\theta < \infty.$$

We denote the space of such functions by $L^2(S^1)$. We state (without proof) that a basis for this space is given by:

$$\{1, \cos(n\theta), \sin(n\theta), \quad n = 1, \ldots\}$$

Hence, an arbitrary function $\psi(\theta) \in L^2(S^1)$ can be expressed as:

$$\psi(\theta) = \frac{a_0}{2} + \sum_{n=1}^{\infty} (a_n \cos(n\theta) + b_n \sin(n\theta)).$$

In discussing the convergence of sequences, the notion of a Cauchy sequence is fundamental.

**Definition 8 (Cauchy Sequence).** A sequence of vectors $\{\psi_n\}$ is said to be a Cauchy sequence if $\forall \epsilon > 0 \; \exists N \in \mathbb{N}$ such that for all $m, n > N$

$$\|\psi_m - \psi_n\| < \epsilon.$$

We now are at the point where we can define the notion of a *Hilbert space*.

**Definition 9 (Hilbert Space).** A vector space, $V$ with an inner product that is "complete" (i.e. every Cauchy sequence converges strongly to a vector in $V$) is called a Hilbert space.

We now return to the notion of a spanning set and basis that incorporates the inner product.

**Definition 10 (Spanning Set, Basis).** A set of vectors $S$ spans an inner product space $V$ if there is no other non-zero vector $\psi \in V$, with $(\psi, \phi) = 0$, $\forall \phi \in S$. It forms a basis if no vector can be removed from $S$ without changing the span.

With the mathematical apparatus surrounding complex inner product spaces defined, we can now define *the state space and state vectors of a quantum mechanical system.*

**Definition 11 (State Space and State Vectors of a Quantum Mechanical System).** The state space of a quantum mechanical system is a (complex) Hilbert space. A state vector of a quantum mechanical system is a vector in the state space having unit magnitude.

From the point of view of physics, two state vectors are "the same" if one is equal to the other multiplied by a complex number. In this sense a state vector is often referred to as *a ray in Hilbert space.*[2]

## 1.2  Linear Operators

Linear operators on the state space describe both measurable physical quantities (i.e. "observables") and how state vectors change in time (i.e. "dynamics"). In particular, we have the following definition.

**Definition 12 (Observable).** Observables in quantum mechanics are described by self-adjoint linear operators.

However, we are getting a bit ahead of ourselves with this definition. First, we need to define the notion of "linear operator" and then the notion of a "self-adjoint linear operator".[3]

---

[2]This is a very important point, and is related to "normalization".

[3]We want to address the terminology of "self-adjoint" versus "Hermitian". Mathematicians tend to prefer "self-adjoint" and physicists prefer "Hermitian". Do they mean the same thing? For finite dimensional, complex vector spaces equipped with an inner product there is no difference in what these two terms mean. In infinite dimensions, however, there are "subtle" differences depending on issues related to boundedness versus unboundedness of the operators, and the nature of the domains on which the operators are defined (which may be dictated by boundary conditions). This is something that you should be aware of, but it is not necessary for us to go into such details here. More details about this issue, including a number of enlightening examples, can be found in the paper of Gieres

François Gieres. Mathematical surprises and Dirac's formalism in quantum mechanics. *Reports on Progress in Physics*, 63(12):1893, 2000.

**Definition 13 (Linear Operator).** A linear operator $A$ on a vector space $V$ assigns to vectors $\psi \in V$ a vector $A\psi \in V$ such that

$$A\left(\alpha\psi + \beta\phi\right) = \alpha A\psi + \beta A\phi, \quad \forall \alpha,\, \beta \in \mathbb{C},\, \psi,\, \phi \in V.$$

We now consider some examples of linear operators.

*Example.* Consider an appropriately defined (i.e. it can be done, but we are not going to work out the details at this point) vector space of complex valued differentiable functions of a real, scalar variable $x$. Then a linear operator on this vector space can be defined by differentiation as follows:

$$(A\psi)(x) = -i\frac{d\psi}{dx}(x).$$

We leave it as an exercise to show that this operator satisfies the condition of linearity given in Definition 13.

*Example.* Consider the complex vector space $\mathbb{C}^3$, with basis $\{\mathbf{e}_1, \mathbf{e}_2, \mathbf{e}_3\}$. Recall from linear algebra (and this is a very important point to recall) that a linear operator on a vector space is defined through the action of the linear operator on each basis element.[4] In particular, for this example suppose that we have:

$$A\mathbf{e}_1 = 3\mathbf{e}_1 + 2i\mathbf{e}_2,$$

$$A\mathbf{e}_2 = -2i\mathbf{e}_1 - \mathbf{e}_2,$$

$$A\mathbf{e}_3 = \mathbf{e}_3. \tag{1.17}$$

More generally, suppose that $\{\mathbf{e}_1, \ldots, \mathbf{e}_n\}$ is a basis of $\mathbb{C}^n$, and let $A : \mathbb{C}^n \to \mathbb{C}^n$ be a linear operator. Now $A\mathbf{e}_i$ is a vector in $\mathbb{C}^n$, for each $i = 1, \ldots, n$. Therefore, it can be represented as a linear combination of the basis elements as follows:

$$A\mathbf{e}_i = \sum_{j=1}^{n} A_{ji}\mathbf{e}_j, \tag{1.18}$$

where $\{A_{ij}\}$ (note the reversal of $i$ and $j$) is the matrix representation of $A$ with respect to the basis $\{\mathbf{e}_1, \ldots, \mathbf{e}_n\}$.

---

[4]This is an important remark that lies at the foundations of much of what we would like to achieve in quantum mechanics. The judicious (and sometimes fortuitous) choice of a basis is central to our ability to understand and solve quantum mechanical problems. We will see this in numerous situations throughout this book.

It should be familiar that the matrix representation of an operator (with respect to a chosen basis) can be used to transform the components of a vector that is represented in the same basis. We demonstrate this fact.

Let

$$\mathbf{x} = \sum_{i=1}^{n} x_i \mathbf{e}_i \quad \text{and} \quad \mathbf{y} = A\mathbf{x} = \sum_{j=1}^{n} y_j \mathbf{e}_j.$$

Then, using (1.18), we can express $\mathbf{y} = A\mathbf{x}$ as:

$$\mathbf{y} = \sum_{i=1}^{n} x_i A\mathbf{e}_i = \sum_{i,j=1}^{n} x_i A_{ji} \mathbf{e}_j,$$

from which it follows that[5]:

$$y_j = \sum_{i=1}^{n} A_{ji} x_i. \tag{1.19}$$

Note that if $\{\mathbf{e}_i\}$ is an orthonormal basis then we have:

$$(\mathbf{e}_j, A\mathbf{e}_i) = \left(\mathbf{e}_j, \sum_{k=1}^{n} A_{ki}\mathbf{e}_k\right) = \sum_{k=1}^{n} A_{ki}(\mathbf{e}_j, \mathbf{e}_k) = \sum_{k=1}^{n} A_{ki}\delta_{jk} = A_{ji}.$$

*Example.* Suppose $\{\mathbf{e}_1, \mathbf{e}_2\}$ is an orthonormal basis of $\mathbb{C}^2$ and let $A : \mathbb{C}^2 \to \mathbb{C}^2$ be a linear operator. Suppose we have:

$$A\mathbf{e}_1 = 3\mathbf{e}_1 + 2i\mathbf{e}_2,$$
$$A\mathbf{e}_2 = \mathbf{e}_2. \tag{1.20}$$

Then the matrix representation of $A$ with respect to the basis $\{\mathbf{e}_1, \mathbf{e}_2\}$ is given by:

$$A = \begin{pmatrix} (\mathbf{e}_1, A\mathbf{e}_1) & (\mathbf{e}_1, A\mathbf{e}_2) \\ (\mathbf{e}_2, A\mathbf{e}_1) & (\mathbf{e}_2, A\mathbf{e}_2) \end{pmatrix} = \begin{pmatrix} 3 & 0 \\ 2i & 1 \end{pmatrix}.$$

Now let

$$\mathbf{x} = 2\mathbf{e}_1 + i\mathbf{e}_2,$$

and let us choose the following explicit representations for $\mathbf{e}_1$ and $\mathbf{e}_2$:

$$\mathbf{e}_1 \to \begin{pmatrix} 1 \\ 0 \end{pmatrix} \qquad \mathbf{e}_2 \to \begin{pmatrix} 0 \\ 1 \end{pmatrix}.$$

---

[5] Compare (1.18) and (1.19). This shows that basis vectors transform differently than the components of vectors. This is important to keep in mind when computing matrix representations of linear operators with respect to a basis.

Then

$$\mathbf{x} = \begin{pmatrix} 2 \\ i \end{pmatrix},$$

and

$$\mathbf{y} = A\mathbf{x} = \begin{pmatrix} 3 & 0 \\ 2i & 1 \end{pmatrix} \begin{pmatrix} 2 \\ i \end{pmatrix} = \begin{pmatrix} 6 \\ 5i \end{pmatrix} = 6\mathbf{e}_1 + 5i\mathbf{e}_2.$$

It is important to keep in mind that the matrix representation of a linear operator depends on the chosen basis, i.e. for two bases $\{\mathbf{e}_i\}$ and $\{\mathbf{f}_i\}$, in general $(\mathbf{f}_i, A\mathbf{f}_j) \neq (\mathbf{e}_i, A\mathbf{e}_j)$.

The following idea will be useful when we define the adjoint of an operator in the next section.

**Definition 14 (Bounded Linear Operator).** A linear operator $A$ defined on $V$ is bounded if there is a positive number $d$ such that $\|A\psi\| \leq d\|\psi\|$ for every $\psi \in V$. The smallest number with this property is called the operator norm of $A$ and is denoted by $\|A\|$.

Every linear operator on a *finite dimensional* Hilbert space is bounded, as we will now show. Let $A$ be a linear operator defined on an $n$ dimensional vector space $V$ that is equipped with an orthonormal basis $\{e_1, \ldots, e_n\}$. Let $A_{ij} \equiv (e_i, Ae_j)$ and let $b$ be the largest of the numbers $|\sum_{i=1}^{n} \bar{A}_{ij} A_{ik}|$ for $j, k = 1, \ldots, n$. Then for any $\mathbf{x} \in V$ we set $\mathbf{y} = A\mathbf{x}$, and consider the following calculation:

$$\|A\mathbf{x}\| = \sum_{i=1}^{n} \bar{y}_i y_i,$$

$$= \sum_{i=1}^{n} \sum_{k=1}^{n} \bar{A}_{ik} \bar{x}_k \sum_{j=1}^{n} A_{ij} x_j,$$

$$= \sum_{j,k=1}^{n} \bar{x}_k \left( \sum_{i=1}^{n} \bar{A}_{ik} A_{ij} \right) x_j,$$

$$\leq \sum_{j,k=1}^{n} |\bar{x}_k| \left| \sum_{i=1}^{n} \bar{A}_{ik} A_{ij} \right| |x_j|,$$

$$\leq b \left( \sum_{k=1}^{n} |x_k|^2 \right),$$

$$\leq b \max_k |x_k|^2 \, n^2,$$

$$\leq bn^2 \sum_{k=1}^{n} |x_k|^2,$$

$$= bn^2 \|\mathbf{x}\|. \tag{1.21}$$

## 1.3  Self-Adjoint Operators, Eigenvalues and Eigenvectors

As we noted earlier, self-adjoint operators[6] play a central role since they are
the mathematical manifestation of "observables" in quantum mechanics. In
this section we will discuss self-adjoint operators in more detail. We begin
with the definition.

**Definition 15 (Adjoint of an Operator).** Let $V$ be a vector space
equipped with an inner product and suppose $A : V \to V$ is a bounded
linear operator. The adjoint of $A$, denoted by $A^\dagger$, is defined by:

$$(\phi, A^\dagger \psi) = (A\phi, \psi), \tag{1.22}$$

for all $\phi, \psi \in V$.

With $A$ known, we can view (1.22) as an equation for the "unknown" $A^\dagger$.
Hence, the question arises, does (1.22) have a solution, i.e. does the adjoint
of a bounded linear operator exist? The answer is "yes", but answering this
question is beyond the scope of this book. You will learn more about the
issues in a course on functional analysis.

In many situations, when we work with the adjoint of a linear operator
on a finite dimensional (complex) vector space, $V$, we will work with its
matrix representation. Here we derive an important property of the matrix
elements.

Using (1.22) and the properties of the (complex) inner product, we have:

$$(\phi, A^\dagger \psi) = (A\phi, \psi) = \overline{(\psi, A\phi)}. \tag{1.23}$$

Now let $\{\mathbf{e}_1, \ldots, \mathbf{e}_n\}$ be an orthonormal basis of $V$. Using (1.23), we have:

$$A^\dagger_{jk} = (\mathbf{e}_j, A^\dagger \mathbf{e_k}) = \overline{(\mathbf{e}_k, A\mathbf{e}_j)} = \overline{A}_{kj}. \tag{1.24}$$

Hence, given a matrix $A$, to compute the adjoint of $A$ we take the transpose
of $A$ and the complex conjugate of each matrix element.

---

[6]When we use the word "operator" in this book we will always mean "linear operator".

If $A$ is a bounded linear operator with a bounded inverse, $A^{-1}$, then $A^\dagger$ has an inverse given by $(A^\dagger)^{-1} = (A^{-1})^\dagger$. This follows from the following calculation and the definition of the adjoint given in Definition 15:

$$(\phi, (A^{-1})^\dagger \psi) = (A^{-1}\phi, \psi) = (A^{-1}\phi, A^\dagger (A^\dagger)^{-1}\psi),$$

$$= (AA^{-1}\phi, (A^\dagger)^{-1}\psi) = (\phi, (A^\dagger)^{-1}\psi).$$

Since this calculation holds for all $\phi$, $\psi$, it follows that:

$$(A^{-1})^\dagger = (A^\dagger)^{-1}.$$

**Definition 16 (Self-adjoint, or Hermitian Operator).** A bounded linear operator, $A$, defined on vector space $V$ equipped with an inner product is said to be self-adjoint, or Hermitian, if

$$(\psi, A\phi) = (A\psi, \phi), \qquad \forall \psi, \phi \in V. \tag{1.25}$$

*Example.* Let $\{\mathbf{e}_1, \mathbf{e}_2\}$ be an orthonormal basis defined on $\mathbb{C}^2$. We define a linear operator on $\mathbb{C}^2$ as follows:

$$
\begin{aligned}
A\mathbf{e}_1 &= 3\mathbf{e}_1 + 2i\mathbf{e}_2, \\
A\mathbf{e}_2 &= -2i\mathbf{e}_1 + \mathbf{e}_2.
\end{aligned}
\tag{1.26}
$$

The matrix representation of $A$ with respect to the basis $\{\mathbf{e}_1, \mathbf{e}_2\}$ is given by:

$$A = \begin{pmatrix} 3 & -2i \\ 2i & 1 \end{pmatrix} = \overline{A^T} = A^\dagger,$$

and we see that $A$ is self-adjoint.

A useful identity is the following. Let $A$ and $B$ be bounded linear operators on a complex vector space equipped with an inner product. Then we have:

$$(AB)^\dagger = B^\dagger A^\dagger.$$

The proof of this identity is left as an exercise.

The notion of eigenvalues and eigenvectors plays a central role in both the mathematical structure and the physical interpretation of quantum mechanics.

**Definition 17 (Eigenvector, Eigenvalue).** $\psi$ is an eigenvector of an operator $A$ with eigenvalue $\lambda$ if

$$A\psi = \lambda\psi, \quad \lambda \in \mathbb{C}.$$

Recall from Definition 11 that the state space of a quantum mechanical system is a complex vector space and the state vector is a vector in the state space having unit magnitude. When the state vector is an eigenvector it is often referred to as an *eigenstate*.

The position and momentum operators in three dimensions are fundamental operators in quantum mechanics.

**Definition 18 (The Position and Momentum Operators in Three Dimensions).** The position operator, $X$, in three dimensions has components $X_j$, $j = 1, 2, 3$, that are defined on functions $\psi$ in a Hilbert space $\mathcal{H}$ by:

$$(X_j \psi)(\mathbf{x}) = x_j \psi(\mathbf{x}). \tag{1.27}$$

The momentum operator $P$ in three dimensions has components $P_j$, $j = 1, 2, 3$ that are defined on differentiable $\psi$ in a Hilbert space $\mathcal{H}$ by

$$(P_j \psi)(\mathbf{x}) = \frac{\hbar}{i} \frac{\partial}{\partial x_j} \psi(\mathbf{x}). \tag{1.28}$$

*One Space Dimension.* In one dimension the position and momentum operators have particularly simple forms:

$$(X\psi)(x) = x\psi(x),$$
$$P\psi(x) = \frac{\hbar}{i} \frac{d\psi}{dx}. \tag{1.29}$$

A Hilbert space that will be useful for describing one dimensional quantum mechanical systems is the space of functions defined on $\mathbb{R}$ (with particular functions denoted by $\psi(x)$) satisfying:

$$\int_{\mathbb{R}} |\psi(x)|^2 dx < \infty.$$

In other words, $X$ and $P$ are defined only for those functions $\psi(x)$ such that:

$$\|P\psi\|^2 = \int_{\mathbb{R}} \left| \frac{\hbar}{i} \frac{d\psi}{dx} \right|^2 dx < \infty,$$
$$\|X\psi\|^2 = \int_{\mathbb{R}} x^2 |\psi(x)|^2 dx < \infty. \tag{1.30}$$

It is a simple calculation to show that the position and momentum operators are linear operators.[7]

The one dimensional position and momentum operators illustrate the issues that occur with unbounded operators. In particular, on the space $L^2(\mathbb{R})$ of square integrable functions $X$ and $P$ are not bounded. They are not defined for all functions in $L^2(\mathbb{R})$, but only on a subset of this Hilbert space. The problem of determining whether or not an unbounded operator is self-adjoint is complicated by the subset of the Hilbert space on which they are defined. On the subsets on which they are defined, these operators are self-adjoint. This can be understood from the following calculations:

$$(X\phi, \psi) = \int_{\mathbb{R}} \bar{x}\overline{\phi(x)}\psi(x)dx = \int_{\mathbb{R}} \overline{\phi(x)}x\psi(x)dx = (\phi, X\psi),$$

$$(1.31)$$

and

$$(P\phi, \psi) = \int_{\mathbb{R}} \overline{\left(\frac{\hbar}{i}\frac{d\phi}{dx}\right)}\psi dx,$$

$$= -\frac{\hbar}{i}\int_{\mathbb{R}} \frac{d\bar{\phi}}{dx}\psi dx,$$

$$= -\frac{\hbar}{i}\left([\bar{\phi}\psi]_{-\infty}^{\infty} - \int_{\mathbb{R}} \bar{\phi}\frac{d\psi}{dx}dx\right),$$

$$= -\frac{\hbar}{i}[\bar{\phi}\psi]_{-\infty}^{\infty} + \int_{\mathbb{R}} \bar{\phi}\frac{\hbar}{i}\frac{d\psi}{dx}dx,$$

$$= (\phi, P\psi). \qquad (1.32)$$

These formal calculations "make sense" for functions $\chi(x) \in L^2(\mathbb{R})$ having the property that:

$$\lim_{x\to\pm\infty} \chi(x) = 0.$$

---

[7]The proof that the one dimensional position operator is linear is straightforward. For appropriate functions, $\psi(x)$, $\phi(x)$ and complex numbers $\alpha$, $\beta \in \mathbb{C}$, we have:

$$X(\alpha\psi(x) + \beta\phi(x))$$

$$= x(\alpha\psi(x) + \beta\phi(x)),$$

$$= x\alpha\psi(x) + x\beta\phi(x),$$

$$= \alpha x\psi(x) + \beta x\phi(x),$$

$$= \alpha X\psi(x) + \beta X\phi(x).$$

The proof that the one dimensional momentum operator is linear follows from the fact that differentiation is a linear operation.

Next we define another common operator in quantum mechanics using the position and momentum operators that we have just defined.

**Definition 19 (Hamiltonian Operator).** For twice differentiable functions $\psi(\mathbf{x})$ the Hamiltonian operator is defined as:

$$(H\psi)(\mathbf{x}) = -\frac{\hbar^2}{2m}\nabla^2\psi(\mathbf{x}) + V(\mathbf{x})\psi(\mathbf{x}), \tag{1.33}$$

where $V(\mathbf{x})$ is the potential energy function. In terms of the position and momentum operators defined in (1.27) and (1.28), respectively, (1.33) can be re-written as:

$$H = \frac{1}{2m}P^2 + V(X). \tag{1.34}$$

It is easy to see that (1.34) is self-adjoint from the following calculation:

$$H^\dagger = \left(\frac{1}{2m}P^2 + V(X)\right)^\dagger,$$

$$= \frac{1}{2m}(P^2)^\dagger + (V(X))^\dagger,$$

$$= \frac{1}{2m}(P^\dagger)^2 + V(X^\dagger),$$

$$= \frac{1}{2m}(P)^2 + V(X) = H.$$

The equality, $(V(X))^\dagger = V(X^\dagger)$, which is essential for the validity of the result, is taken as an *assumption*. It will be straightforward to verify that the potential energy functions considered in this book satisfy this requirement. It is an interesting mathematics problem to determine conditions on potential energy functions for which this equality holds, but that is beyond the scope of this book.[8]

---

[8] In this definition we see a procedure for associating a quantum mechanical Hamiltonian operator with a classical Hamiltonian function. In particular, in the classical Hamiltonian function we merely replace the momentum and position variables by their quantum mechanical operator counterparts. This method of canonical quantization starting from the classical Hamiltonian framework was pioneered by Dirac.

P. A. M. Dirac. *The principles of quantum mechanics*. Number 27. Oxford university press, 1981.

## 1.4  Dirac Notation

In this section we will introduce a notation, proposed by Dirac,[9] that renders many of the conceptual issues associated with computing quantum mechanical quantities in linear vector spaces simple and transparent. Throughout the rest of the book we will, almost exclusively, use *Dirac notation*, which we now describe.[10]

Let $V$ be a complex vector space equipped with an inner product. Rather than denote vectors in $V$ by $\psi$, we will denote them by

$$|\psi\rangle \in V, \tag{1.35}$$

which we refer to as a "ket", or "ket vector". Hence, the state vector of a quantum mechanical system is given by a normalised ket.

As an example, we express (1.26) in Dirac notation. In particular, we consider the complex vector space $\mathbb{C}^2$. We denote an orthonormal basis on $\mathbb{C}^2$ by $\{\mathbf{e}_1, \mathbf{e}_2\}$, or $\{|\mathbf{e}_1\rangle, |\mathbf{e}_2\rangle\}$. We can define a linear operator, $A$, on $\mathbb{C}^2$ by defining it's action on the basis vectors as follows[11]:

$$A|\mathbf{e}_1\rangle = 3|\mathbf{e}_1\rangle + 2i|\mathbf{e}_2\rangle,$$
$$A|\mathbf{e}_2\rangle = -2i|\mathbf{e}_1\rangle + |\mathbf{e}_2\rangle. \tag{1.36}$$

For $|\phi\rangle$, $|\psi\rangle \in V$ we denote their inner product by:

$$\langle\phi|\psi\rangle \equiv (|\phi\rangle, |\psi\rangle),$$

$$\text{(where the previous notation was } (\phi, \psi)), \tag{1.37}$$

where $\langle\phi|$ is referred to as a "bra". Hence the inner product pairing defined by (1.37) is referred to as a "bra-ket" (a misspelling of "bracket").

*Example.* Consider the complex vector space $\mathbb{C}^2$ and let $\{|\mathbf{e}_1\rangle, |\mathbf{e}_2\rangle\}$ denote an orthonormal basis on $\mathbb{C}^2$, i.e. $\langle\mathbf{e}_i|\mathbf{e}_j\rangle = \delta_{ij}$. Then arbitrary vectors

---

[9]P. A. M. Dirac. A new notation for quantum mechanics. In *Mathematical Proceedings of the Cambridge Philosophical Society*, volume 35, pages 416–418. Cambridge University Press, 1939.

[10]Computations involving the inner product and the adjoint play an important role in quantum mechanics. Dirac notation serves to simplify the set-up for these computations in a way that makes them almost obvious. You should get a sense of that in this section, but it should become particularly apparent in Chapter 3.

[11]Note that this example is just a reformulation of (1.26) in Dirac notation.

$|\psi\rangle, |\phi\rangle \in \mathbb{C}^2$ can be written as:

$$|\psi\rangle = a|\mathbf{e}_1\rangle + b|\mathbf{e}_2\rangle,$$
$$|\phi\rangle = c|\mathbf{e}_1\rangle + d|\mathbf{e}_2\rangle, \tag{1.38}$$

and we have:

$$
\begin{aligned}
\langle\phi|\psi\rangle = (|\phi\rangle, |\psi\rangle) &= (c|\mathbf{e}_1\rangle + d|\mathbf{e}_2\rangle, a|\mathbf{e}_1\rangle + b|\mathbf{e}_2\rangle), \\
&= a\,(c|\mathbf{e}_1\rangle + d|\mathbf{e}_2\rangle, |\mathbf{e}_1\rangle) \\
&\quad + b\,(c|\mathbf{e}_1\rangle + d|\mathbf{e}_2\rangle, |\mathbf{e}_2\rangle), \\
&= a\bar{c}\,(|\mathbf{e}_1\rangle, |\mathbf{e}_1\rangle) + a\bar{d}\,(|\mathbf{e}_2\rangle, |\mathbf{e}_1\rangle) \\
&\quad + b\bar{c}\,(|\mathbf{e}_1\rangle, |\mathbf{e}_2\rangle) + b\bar{d}\,(|\mathbf{e}_2\rangle, |\mathbf{e}_2\rangle), \\
&= a\bar{c}\,\langle\mathbf{e}_1|\mathbf{e}_1\rangle + a\bar{d}\,\langle\mathbf{e}_2|\mathbf{e}_1\rangle \\
&\quad + b\bar{c}\,\langle\mathbf{e}_1|\mathbf{e}_2\rangle + b\bar{d}\,\langle\mathbf{e}_2|\mathbf{e}_2\rangle, \\
&= a\bar{c} + b\bar{d} = (\bar{c}\ \bar{d}) \begin{pmatrix} a \\ b \end{pmatrix}
\end{aligned}
\tag{1.39}
$$

With respect to the basis $\{|\mathbf{e}_1\rangle, |\mathbf{e}_2\rangle\}$ we can make the following identifications that should be familiar from your experience with matrix algebra:

$$|\phi\rangle \leftarrow \begin{pmatrix} c \\ d \end{pmatrix}, \tag{1.40}$$

$$\langle\phi| \leftarrow (\bar{c}\ \bar{d}). \tag{1.41}$$

In the context of this example, we can examine linear operators and Dirac notation. Let $A$ be a linear operator on $\mathbb{C}^2$. We consider $A|\psi\rangle$ and the inner product of this vector with another vector $|\phi\rangle$:

$$(|\phi\rangle, A|\psi\rangle) = \langle\phi|A|\psi\rangle. \tag{1.42}$$

The equation (1.42) is referred to as the *expectation value of $A$ in the state* $|\psi\rangle$ and we will see this idea many times throughout this book. Note that we have:

$$A_{ij} = \langle\mathbf{e}_i|A|\mathbf{e}_j\rangle, \text{ which is the same as } (\mathbf{e}_i.A\mathbf{e}_j), \ (|\mathbf{e}_i\rangle, A|\mathbf{e}_j\rangle).$$

We derived this expression for the matrix elements of a linear operator with respect to a given basis earlier. Here we show how the same relation holds in Dirac notation. We can express the action of $A$ on each basis element as follows:

$$A|\mathbf{e}_j\rangle = \sum_k A_{kj}|\mathbf{e}_k\rangle.$$

Using this relation, it follows that:

$$\langle\mathbf{e}_i|A|\mathbf{e}_j\rangle = \sum_k A_{kj}\langle\mathbf{e}_i|\mathbf{e}_k\rangle = \sum_k A_{kj}\delta_{ik} = A_{ij}. \tag{1.43}$$

As noted earlier, any $|\phi\rangle$, $|\psi\rangle \in \mathbb{C}^2$ can be expressed as:

$$\begin{aligned}
|\phi\rangle &= c|\mathbf{e}_1\rangle + d|\mathbf{e}_2\rangle, \\
|\psi\rangle &= a|\mathbf{e}_1\rangle + b|\mathbf{e}_2\rangle.
\end{aligned} \tag{1.44}$$

Then if:

$$A = \begin{pmatrix} a_{11} & a_{12} \\ a_{21} & a_{22} \end{pmatrix}, \tag{1.45}$$

we have:

$$\langle\phi|A|\psi\rangle = (\bar{c}\,\bar{d}) \begin{pmatrix} a_{11} & a_{12} \\ a_{21} & a_{22} \end{pmatrix} \begin{pmatrix} a \\ b \end{pmatrix} \tag{1.46}$$

*Operators in Dirac Notation.* Let $V$ be a complex vector space equipped with an inner product, and consider the "kets" $|u\rangle$, $|v\rangle$, $|w\rangle \in V$. Using a bra and a ket, we define a linear operator as follows:

$$B = |u\rangle\langle v|.$$

$B$ acts on kets, $|w\rangle \in V$, as follows:

$$B|w\rangle = |u\rangle \underbrace{\langle v|w\rangle}_{\text{complex number}} = \underbrace{\langle v|w\rangle}_{\text{complex number}} |u\rangle.$$

In particular, let $V = \mathbb{C}^n$ and let $\{\mathbf{e}_1, \ldots, \mathbf{e}_n\}$ be an orthonormal basis of $\mathbb{C}^n$, i.e. $\langle\mathbf{e}_i|\mathbf{e}_j\rangle = \delta_{ij}$. Then define $B$ as follows:

$$B = |\mathbf{e}_2\rangle\langle\mathbf{e}_1|.$$

Then we have:

$$\begin{aligned}
B|\mathbf{e}_1\rangle &= |\mathbf{e}_2\rangle\langle\mathbf{e}_1|\mathbf{e}_1\rangle = |\mathbf{e}_2\rangle, \\
B|\mathbf{e}_2\rangle &= |\mathbf{e}_2\rangle\langle\mathbf{e}_1|\mathbf{e}_2\rangle = |\mathbf{e}_2\rangle \cdot 0 = 0.
\end{aligned} \tag{1.47}$$

Since the basis is orthonormal we can represent an arbitrary linear operator on $\mathbb{C}^n$ as follows:

$$A = \sum_{ij} A_{ij} |\mathbf{e}_i\rangle\langle\mathbf{e}_j|, \tag{1.48}$$

and using the fact that a linear operator can be defined by specifying how it acts on each basis element. In this case, it is as follows:

$$A|\mathbf{e}_k\rangle = \sum_{ij} A_{ij} |\mathbf{e}_i\rangle\langle\mathbf{e}_j|\mathbf{e}_k\rangle,$$

$$= \sum_{ij} A_{ij} |\mathbf{e}_i\rangle\delta_{jk},$$

$$= \sum_{i} A_{ik} |\mathbf{e}_i\rangle. \tag{1.49}$$

In particular, we consider the special case of $\mathbb{C}^2$. In this case we specify a linear operator acting on each basis element in the following manner:

$$A|\mathbf{e}_1\rangle = a|\mathbf{e}_1\rangle + c|\mathbf{e}_2\rangle,$$
$$A|\mathbf{e}_2\rangle = b|\mathbf{e}_1\rangle + d|\mathbf{e}_2\rangle, \tag{1.50}$$

from which it follows that the matrix representation of the operator with respect to this particular basis is given by:

$$A \leftrightarrow \begin{pmatrix} a & b \\ c & d \end{pmatrix}.$$

From (1.49), we have:

$$A = a|\mathbf{e}_1\rangle\langle\mathbf{e}_1| + b|\mathbf{e}_1\rangle\langle\mathbf{e}_2| + c|\mathbf{e}_2\rangle\langle\mathbf{e}_1| + d|\mathbf{e}_2\rangle\langle\mathbf{e}_2|. \tag{1.51}$$

This expression can be verified by considering the action of $A$ in this representation on each basis element:

$$A|\mathbf{e}_1\rangle = a|\mathbf{e}_1\rangle\langle\mathbf{e}_1|\mathbf{e}_1\rangle + b|\mathbf{e}_1\rangle\langle\mathbf{e}_2|\mathbf{e}_1\rangle$$

$$+ c|\mathbf{e}_2\rangle\langle\mathbf{e}_1|\mathbf{e}_1\rangle + d|\mathbf{e}_2\rangle\langle\mathbf{e}_2|\mathbf{e}_1\rangle,$$

$$= a|\mathbf{e}_1\rangle + c|\mathbf{e}_2\rangle. \tag{1.52}$$

Similarly,

$$A|\mathbf{e}_2\rangle = b|\mathbf{e}_1\rangle + d|\mathbf{e}_2\rangle. \tag{1.53}$$

*Adjoints in Dirac Notation.* Recall that we showed earlier that given the matrix representation of a linear operator, the matrix representation of the adjoint of the linear operator (with respect to the same basis) is obtained by taking the transpose of the matrix, and then the complex conjugate of each matrix element. For example, if we consider a linear operator $A$ on a two dimensional complex vector space and its matrix representation with respect to some basis:

$$A \leftrightarrow \begin{pmatrix} a & b \\ c & d \end{pmatrix}, \tag{1.54}$$

then the matrix representation of the adjoint of $A$ with respect to the same basis is given by:

$$A^\dagger \leftrightarrow \begin{pmatrix} \bar{a} & \bar{c} \\ \bar{b} & \bar{d} \end{pmatrix}. \tag{1.55}$$

We can now see how this fact would be manifested in Dirac notation. Consider

$$A = a|\mathbf{e}_1\rangle\langle\mathbf{e}_1| + b|\mathbf{e}_1\rangle\langle\mathbf{e}_2| + c|\mathbf{e}_2\rangle\langle\mathbf{e}_1| + d|\mathbf{e}_2\rangle\langle\mathbf{e}_2|, \tag{1.56}$$

then in order for the matrix representation to agree with (1.55) we must have:

$$A^\dagger = \bar{a}|\mathbf{e}_1\rangle\langle\mathbf{e}_1| + \bar{b}|\mathbf{e}_2\rangle\langle\mathbf{e}_1| + \bar{c}|\mathbf{e}_1\rangle\langle\mathbf{e}_2| + \bar{d}|\mathbf{e}_2\rangle\langle\mathbf{e}_2|. \tag{1.57}$$

Note how each term in (1.56) transforms under the adjoint operation. For example, if we have $A = a|\psi\rangle\langle\phi|$, then:

$$A^\dagger = (a|\psi\rangle\langle\phi|)^\dagger = \bar{a}|\phi\rangle\langle\psi|.$$

We can also see this from the following calculation:

$$\langle u|A^\dagger|v\rangle = \langle Au|v\rangle,$$

(definition of the adjoint in Dirac notation),

$$= \overline{\langle v|A|u\rangle},$$

(property of the inner product in Dirac notation),

$$= \overline{\langle v|(a|\psi\rangle\langle\phi|)|u\rangle},$$

$$= \overline{\bar{a}\langle v|\psi\rangle\langle\phi|u\rangle},$$

$$= \bar{a}\overline{\langle v|\psi\rangle}\,\overline{\langle\phi|u\rangle},$$

$$= \bar{a}\langle\psi|v\rangle\langle u|\phi\rangle,$$

$$= \bar{a}\langle u|\phi\rangle\langle\psi|v\rangle,$$

$$= \langle u|(\bar{a}|\phi\rangle\langle\psi|)|v\rangle, \tag{1.58}$$

and since this is true for any $|u\rangle$, $|v\rangle$ we can conclude that:

$$A^\dagger = (a|\psi\rangle\langle\phi|)^\dagger = \bar{a}|\phi\rangle\langle\psi|.$$

In particular, if we apply this result to the operator:

$$|\mathbf{e}_1\rangle\langle\mathbf{e}_2| + |\mathbf{e}_2\rangle\langle\mathbf{e}_1|,$$

we see that this operator is self-adjoint (you will need to use the fact that the "adjoint of the sum of two operators is the sum of the adjoint of each operator", but you showed that in the exercises).

Moreover, this calculation is also consistent with the following identity:

$$(|\psi\rangle)^\dagger = \langle\psi|.$$

Now we want to determine the bra associated with the ket $A|\psi\rangle$. Towards this end, we let $|u\rangle = A|\psi\rangle$, and then we perform the following calculation:

$$\langle u|v\rangle = \overline{\langle v|u\rangle},$$

$$= \overline{\langle v|A|\psi\rangle},$$

$$= \langle\psi|A^\dagger|v\rangle, \tag{1.59}$$

which is true for any $|v\rangle$. Hence, we have:

$$\langle u| = (A|\psi\rangle)^\dagger = \langle\psi|A^\dagger. \tag{1.60}$$

## 1.5 Projection Operators and the Spectral Theorem

Next we define the notion of a projection operator. These play an important role in quantum mechanics, especially with respect to the notion of *measurement*.

**Definition 20 (Projection Operator).** Let $P : V \to V$ be a linear operator on a vector space $V$. Then $P$ is said to be a projection operator if $P^2 = P$.

In Dirac notation an example of a projection operator would be a "bra-ket combination" of the form $|\phi\rangle\langle\phi|$, where the ket $|\phi\rangle$ is normalised to have norm one. Note that projection operators of this form are *self-adjoint*.

*Example.* Let $\{|\mathbf{e}_1\rangle, |\mathbf{e}_2\rangle\}$ be an orthonormal basis of $\mathbb{C}^2$. and consider the projection operator defined as $P = |\mathbf{e}_1\rangle\langle\mathbf{e}_1|$ and let it act on $|\psi\rangle = a|\mathbf{e}_1\rangle + b|\mathbf{e}_2\rangle$. Then we have:

$$P(a|\mathbf{e}_1\rangle + b|\mathbf{e}_2\rangle) = (|\mathbf{e}_1\rangle\langle\mathbf{e}_1|)(a|\mathbf{e}_1\rangle + b|\mathbf{e}_2\rangle) = a|\mathbf{e}_1\rangle.$$

Note that we have:

$$P^2 = |\mathbf{e}_1\rangle\langle\mathbf{e}_1|\mathbf{e}_1\rangle\langle\mathbf{e}_1| = |\mathbf{e}_1\rangle\langle\mathbf{e}_1| = P \quad \text{and} \quad P^\dagger = |\mathbf{e}_1\rangle\langle\mathbf{e}_1| = P.$$

More generally, it is easy to see that for an orthonormal basis of $\mathbb{C}^n$:

$$|\mathbf{e}_1\rangle, \ldots, |\mathbf{e}_n\rangle \in \mathbb{C}^n,$$

the operator defined by:

$$\sum_{k=1}^{d} |\mathbf{e}_k\rangle\langle\mathbf{e}_k|,$$

is a self adjoint projection operator onto the subspace of dimension $d$ spanned by $\mathbf{e}_1, \ldots, \mathbf{e}_d$, $d \leq n$.

As another example, consider $\mathbb{C}^3$ with the orthonormal basis $\{|\mathbf{e}_1, |\mathbf{e}_2\rangle, |\mathbf{e}_3\rangle\}$. Then define the self adjoint operator

$$P = |\mathbf{e}_1\rangle\langle\mathbf{e}_1| + |\mathbf{e}_2\rangle\langle\mathbf{e}_2|,$$

and consider it action on an arbitrary vector

$$|\psi\rangle = a|\mathbf{e}_1\rangle + b|\mathbf{e}_2\rangle + c|\mathbf{e}_3\rangle.$$

Doing this calculation, we obtain:

$$P|\psi\rangle = (|\mathbf{e}_1\rangle\langle\mathbf{e}_1| + |\mathbf{e}_2\rangle\langle\mathbf{e}_2|)(a|\mathbf{e}_1\rangle + b|\mathbf{e}_2\rangle + c|\mathbf{e}_3\rangle),$$

$$= a|\mathbf{e}_1\rangle + b|\mathbf{e}_2\rangle, \tag{1.61}$$

which is the projection of $|\psi\rangle$ onto the subspace of $\mathbb{C}^3$ spanned by $|\mathbf{e}_1\rangle$ and $|\mathbf{e}_2\rangle$. Moreover, it should be easy to see that for this example we have: $P^2 = P$ and $P^\dagger = P$.

*Remark.* If $\{|\mathbf{e}_1\rangle, \ldots, |\mathbf{e}_n\rangle\}$ is an orthonormal basis of $\mathbb{C}^n$, then the operator:

$$\mathbb{I} = \sum_{i=1}^{n} |\mathbf{e}_i\rangle\langle\mathbf{e}_i|,$$

is a self adjoint projection operator onto the entire space. This can be seen as follows. Consider an arbitrary vector, $|\psi\rangle \in \mathbb{C}^n$:

$$|\psi\rangle = \sum_{i=1}^{n} a_i |\mathbf{e}_i\rangle.$$

Then we have:

$$\mathbb{I}|\psi\rangle = \sum_{ij} a_j |\mathbf{e}_i\rangle\langle\mathbf{e}_i|\mathbf{e}_j\rangle = \sum_{ij} a_j |\mathbf{e}_i\rangle\delta_{ij} = \sum_{i} a_i |\mathbf{e}_i\rangle = |\psi\rangle. \tag{1.62}$$

The following result describes some extremely important properties of self-adjoint operators that play a fundamental role in quantum mechanics.

**Theorem 1 (Spectral Theorem for Finite Dimensional Self-Adjoint Operators).** *Let $A : V \to V$ be a self-adjoint linear operator on a finite dimensional complex inner product space. Then:*

i) *$A$ has real eigenvalues, and the eigenvectors of $A$ corresponding to distinct eigenvalues are orthogonal.*

ii) *The eigenvectors of $A$ span $V$.*

**Proof.** We begin by showing that $A$ has real eigenvalues. Let $|\mathbf{e}\rangle$ be a normalised eigenvector of $A$ with eigenvalues $\lambda$, i.e.,

$$A|\mathbf{e}\rangle = \lambda|\mathbf{e}\rangle, \qquad \langle\mathbf{e}|\mathbf{e}\rangle = 1.$$

Then we have:

$$\begin{aligned}
\lambda &= \langle\mathbf{e}|A|\mathbf{e}\rangle = \langle A^\dagger\mathbf{e}|\mathbf{e}\rangle, \\
&= \overline{\langle\mathbf{e}|A^\dagger|\mathbf{e}\rangle}, \\
&= \overline{\langle\mathbf{e}|A|\mathbf{e}\rangle}, \quad \text{since} \quad A = A^\dagger, \\
&= \overline{\langle\mathbf{e}|\lambda|\mathbf{e}\rangle}, \\
&= \bar{\lambda}\,\overline{\langle\mathbf{e}|\mathbf{e}\rangle}, \\
&= \bar{\lambda}, \tag{1.63}
\end{aligned}$$

from which it follows that $\lambda$ is real.

Next we show that eigenvectors corresponding to distinct eigenvalues are orthogonal, i.e. we consider eigenvectors $|\mathbf{e}_1\rangle$, $|\mathbf{e}_2\rangle$ such that:

$$\begin{aligned}
A|\mathbf{e}_1\rangle &= \lambda_1|\mathbf{e}_1\rangle, \\
A|\mathbf{e}_2\rangle &= \lambda_2|\mathbf{e}_2\rangle.
\end{aligned} \tag{1.64}$$

Then we have:

$$\begin{aligned}
\lambda_1 \langle \mathbf{e}_2 | \mathbf{e}_1 \rangle &= \langle \mathbf{e}_2 | A | \mathbf{e}_1 \rangle \\
&= \langle A^\dagger \mathbf{e}_2 | \mathbf{e}_1 \rangle \\
&= \overline{\langle \mathbf{e}_1 | A^\dagger | \mathbf{e}_2 \rangle}, \\
&= \overline{\langle \mathbf{e}_1 | A | \mathbf{e}_2 \rangle}, \quad \text{since} \quad A = A^\dagger, \\
&= \bar{\lambda}_2 \overline{\langle \mathbf{e}_1 | \mathbf{e}_2 \rangle}, \\
&= \lambda_2 \langle \mathbf{e}_2 | \mathbf{e}_1 \rangle.
\end{aligned} \tag{1.65}$$

Hence, we have:

$$(\lambda_1 - \lambda_2)\langle \mathbf{e}_2 | \mathbf{e}_1 \rangle = 0 \quad \Rightarrow \quad \langle \mathbf{e}_2 | \mathbf{e}_1 \rangle = 0.$$

Now we return to the final part of this result — "eigenvectors of $A$ span $V$". The issue we need to deal with here is the possibility of two (or more) eigenvectors having the same eigenvalue, i.e. *degeneracy*, since if we have $n$ distinct eigenvalues, we have $n$ distinct eigenvectors, and we have proven above that the eigenvectors are orthogonal.

First, we note that in linear algebra it is proven that a self-adjoint operator on a $n$-dimensional, complex inner produce space has $n$ orthogonal eigenvectors. What is *not* necessarily true is that the eigenvalues corresponding to each of these eigenvectors are all different. What we have shown above is that if we have $n$ distinct eigenvalues, then we have $n$ orthogonal eigenvectors.

We examine this situation more closely assuming that we have $n$ orthogonal eigenvectors. Suppose the eigenvectors $|\mathbf{e}_1\rangle$, $|\mathbf{e}_2\rangle$ have the same eigenvalue, $\lambda$, i.e.

$$\begin{aligned}
A|\mathbf{e}_1\rangle &= \lambda|\mathbf{e}_1\rangle, \\
A|\mathbf{e}_2\rangle &= \lambda|\mathbf{e}_2\rangle.
\end{aligned} \tag{1.66}$$

Then it follows that any linear combination of $|\mathbf{e}_1\rangle$ and $|\mathbf{e}_2\rangle$ is also an eigenvector having eigenvalue $\lambda$:

$$A(\alpha|\mathbf{e}_1\rangle + \beta|\mathbf{e}_2\rangle) = \lambda(\alpha|\mathbf{e}_1\rangle + \beta|\mathbf{e}_2\rangle).$$

This calculation can easily be generalised to the case where more than two eigenvectors have the same eigenvalue. Hence we can conclude that the set of eigenvectors corresponding to the same eigenvalue spans a subspace, $V_\lambda$, of $V$. We let $P_\lambda$ denote the projection onto the subspace $V_\lambda$. It can be proven that $P_\lambda$ is self adjoint. This same procedure can be carried out for each eigenvalue having more than one eigenvector. In this way $V$ is represented as the direct sum of subspaces corresponding to the span of the eigenvectors corresponding to a given eigenvalue. We can also construct self adjoint projection operators corresponding to each subspace. The case where the eigenvalues are distinct is particularly simple. Each of the subspaces in that case are just one dimensional, and the projection operators are easily constructed from the bra and ket associated with each eigenvalue in the manner that we have already described. $\qquad\square$

Before proceeding with the remainder of the proof of this theorem, we make an important remark that uses what we have shown up to this point.

In completing this section we describe a very important idea for representing a self adjoint operator based on the above theorem.

*Spectral representation of A in Dirac notation.* We have shown that if $A$ is a self-adjoint linear operator on $\mathbb{C}^n$ having distinct eigenvalues, then its normalised eigenvectors, denoted $\{|\mathbf{e}_1\rangle, \ldots, |\mathbf{e}_n\rangle\}$ form a basis of $\mathbf{C}^n$. We can therefore write $A$ in the form[12]:

$$A = \sum_{i=1}^{n} \lambda_i |\mathbf{e}_i\rangle\langle \mathbf{e}_i|, \tag{1.67}$$

where $\lambda_i$ is the eigenvalue corresponding to the eigenvector $|\mathbf{e}_i\rangle$ and $|\mathbf{e}_i\rangle\langle \mathbf{e}_i|$ is the projection operator onto the subspace spanned by $|\mathbf{e}_i\rangle$. This can be verified with the following "check" of how $A$ acts on each basis vector (as discussed above):

$$A|\mathbf{e}_k\rangle = \sum_{i=1}^{n} \lambda_i |\mathbf{e}_i\rangle\langle \mathbf{e}_i|\mathbf{e}_k\rangle,$$

$$= \sum_{i=1}^{n} \lambda_i |\mathbf{e}_i\rangle\delta_{ik},$$

$$= \lambda_k |\mathbf{e}_k\rangle. \tag{1.68}$$

In the case where the eigenvalues are not necessarily distinct we can write $A$ in the form:

$$A = \sum_{\alpha} \lambda_\alpha P_\alpha, \tag{1.69}$$

---

[12]Note that the spectral representation of an operator in Dirac notation is the Dirac notation manifestation of the diagonalization of an operator in a basis of eigenvectors.

where $P_\alpha$ is the projection operator onto the subspace of eigenvectors corresponding to the eigenvalue $\lambda_\alpha$. You should convince yourself that (1.69) takes the form (1.67) when the eigenvalues are distinct.

## Problems

1. Triangle inequality, bounded operators.

   (a) Use Schwarz's inequality for the inner product on a Hilbert space to prove the triangle inequality, $\|\psi + \phi\| \leq \|\psi\| + \|\phi\|$.

   (b) If $A$ and $B$ are bounded linear operators with norms $a$ and $b$ respectively, show that $A + B$ and $AB$ are also bounded.

2. Matrix and operator notations.

   Let $\{e_1, e_2\}$ be a basis for $\mathbb{C}^2$ ($e_1$ and $e_2$ need not be normalised or orthogonal). Let $A$ and $B$ be operators on $\mathbb{C}^2$ defined by

$$Ae_1 = e_2; \quad Ae_2 = 2e_1 + e_2$$
$$Be_1 = ie_2; \quad Be_2 = -ie_1. \tag{1.70}$$

   (a) Calculate the matrices $(A)_{ij}$ and $(B)_{ij}$, of $A$ and $B$ with respect to the basis $\{e_1, e_2\}$.

   (b) Calculate $ABe_1$ and $ABe_2$ using (1.70) directly and hence calculate the matrix of $AB$ with respect to the basis $\{e_1, e_2\}$.

   (c) Calculate the matrix of $AB$ by doing matrix multiplication of $(A)_{ij}$ and $(B)_{ij}$ and confirm that your answer agrees with the result of part (b).

   (d) Assume now that $e_1$ and $e_2$ are orthonormal. Let $\phi = 2e_1 + 3ie_2$.

      (i) Use operator notation to calculate $A\phi$ and then the inner product $(\phi, A\phi)$.

      (ii) Calculate the inner product $(\phi, A\phi)$ using matrix notation; confirm that your answer agrees with that in (i).

3. Properties of the adjoint.

   Let $A$ and $B$ be bounded linear operators and $\alpha$ be a complex scalar. Check, using the definition of the adjoint of a bounded linear operator, that

   (a) $(A^\dagger)^\dagger = A$.

   (b) $(A + B)^\dagger = A^\dagger + B^\dagger$.

   (c) $(\alpha A)^\dagger = \bar{\alpha} A^\dagger$.

   (d) $(AB)^\dagger = B^\dagger A^\dagger$.

(e) Assuming that $A$ has a bounded inverse $A^{-1}$, show that $(A^\dagger)^{-1} = (A^{-1})^\dagger$.

4. (a) Use the definition of the adjoint of an operator to show that the adjoint of the operator $|\psi\rangle\langle\phi|$ is $|\phi\rangle\langle\psi|$, where $|\psi\rangle$ and $|\phi\rangle$ are vectors (not necessarily normalised).

   (b) Let $P_k$ be the projection operator $\sum_{i=1}^{k} |e_i\rangle\langle e_i|$, where $\{|e_i\rangle\}$ form an orthonormal basis for $\mathbb{C}^n$, and $k \leq n$.

   Show that $P_k$ is self-adjoint and that $P_k^2 = P_k$.

   (c) Using the result above to show that the eigenvalues of $P_k$ are 1 and 0.

5. Let $H$ be the operator on $\mathbb{C}^2$ defined by

$$H|1\rangle = -\frac{1}{2}|1\rangle + \frac{\sqrt{3}}{2}|2\rangle; \quad H|2\rangle = \frac{\sqrt{3}}{2}|1\rangle + \frac{1}{2}|2\rangle \tag{1.71}$$

where $|1\rangle$ and $|2\rangle$ give an orthonormal basis for $\mathbb{C}^2$.

   (a) Find the matrix of $H$ with respect to the basis $\{|1\rangle, |2\rangle\}$ and show that $H$ is self-adjoint.

   (b) Find the expression for $H$ in Dirac notation in this basis (i.e. write $H$ in the form $\sum_{ij=1}^{2} |i\rangle\langle j|$) and confirm that your expression gives the correct vectors when acting on $|1\rangle$ and $|2\rangle$.

   (c) Show that $H^2 = I$, where $I$ is the identity operator on $\mathbb{C}^2$.

   (d) Calculate the eigenvalues of $H$ and find orthonormal eigenvectors $|e_1\rangle$ and $|e_2\rangle$ for $H$.

   (e) Write $H$ in Dirac notation with respect to the basis of eigenvectors (i.e. write $H$ in the form $\sum_{ij=1}^{2} |e_i\rangle\langle e_j|$) and, by expanding this expression out, confirm that it is equal to the expression you found in (b).

6. Let $H$ be the operator defined in (1.71).

   (a) We define $e^{-itH}$ by

$$e^{-itH} = \sum_{n=0}^{\infty} \frac{(-it)^n H^n}{n!}.$$

   Using $H^2 = I$, or otherwise, write $e^{-itH}$ in the form

$$e^{-itH} = a(t)I + ib(t)H$$

   where $a$ and $b$ are real functions of $t$ which you should find.

   (b) Show that $e^{-itH}$ is unitary.

(c) The state of a quantum system $|\psi(t)\rangle$ at time $t$ is defined by

$$|\psi(t)\rangle = e^{-itH}\,|\psi_0\rangle.$$

Consider the case $|\psi_0\rangle = |1\rangle$. Write $|\psi_0\rangle$ in terms of eigenstates of $H$ and hence, or otherwise, calculate $|\psi(t)\rangle$. Show that $|\psi(t)\rangle$ satisfies Schrödinger's equation (in units in which $\hbar = 1$):

$$i\frac{\partial\,|\psi(t)\rangle}{\partial t} = H\,|\psi(t)\rangle.$$

7. Consider a Hamiltonian defined on $\mathbb{C}^2$ where in some orthonormal basis $\{|v_1\rangle,\ |v_2\rangle\}$ it has the following matrix representation:

$$H = \begin{pmatrix} 3 & 1 \\ 1 & 3 \end{pmatrix}. \tag{1.72}$$

(a) Show that $H$ is self-adjoint.
(b) Express $H$ in Dirac notation in the basis $\{|v_1\rangle\,|v_2\rangle\}$.
(c) Compute the expectation value of $H$ in the state $|v_1\rangle$.
(d) Compute the eigenvalues and the eigenstates of $H$.
(e) Denoting the eigenstates of $H$ by $\{|E_1\rangle,\ |E_2\rangle\}$, express $H$ in this basis using Dirac notation.
(f) Consider the operator:

$$e^{-\frac{iHt}{\hbar}}. \tag{1.73}$$

Express this operator in the basis $\{|E_1\rangle,\ |E_2\rangle\}$ using Dirac notation.

8. Let $|\phi\rangle$, $|\psi\rangle$ denote kets and let $A$ denote a linear operator acting on the kets.

(a) Show that

$$(A|\psi\rangle)^\dagger = \langle\psi|A^\dagger.$$

(b) For the following four expressions state if it is a scalar, ket, bra, or operator and compute its adjoint:

    (i) $\langle\psi|A|\phi\rangle\langle\psi|\phi\rangle$.
    (ii) $\langle\psi|\phi\rangle\langle\psi|A$.
    (iii) $\langle\psi|\phi\rangle A|\phi\rangle\langle\psi|$.
    (iv) $A|\psi\rangle\langle\phi|A|\psi\rangle$.

Now suppose that $A$ is self-adjoint.

(c) Show that the eigenvalues of $A$ are real.

(d) Show that the eigenvectors of $A$ corresponding to distinct eigenvalues are orthogonal.

9. Consider the state space $\mathbb{C}^2$ with orthonormal basis,

$$|1\rangle = \begin{pmatrix} 1 \\ 0 \end{pmatrix}, \quad |2\rangle = \begin{pmatrix} 0 \\ 1 \end{pmatrix}.$$

We define the operators $A$ and $B$ on $\mathbb{C}^2$ with respect to this basis as follows:

$$A|1\rangle = 2|1\rangle - i|2\rangle,$$
$$A|2\rangle = i|1\rangle + 2|2\rangle$$
$$B|1\rangle = |1\rangle - i|2\rangle,$$
$$B|2\rangle = i|1\rangle + |2\rangle$$

(a) Write down the matrix representations of $A$ and $B$ with respect to the basis $|1\rangle$, $|2\rangle$.
(b) Show that $A$ and $B$ are self-adjoint.
(c) Show that $A$ and $B$ commute.
(d) Show that the eigenvalues of $A$ are 1 and 3. Let $|e_1\rangle$ denote the normalized eigenstate corresponding to the eigenvalue 1 and let $|e_2\rangle$ denote the normalized eigenstate corresponding to the eigenvalue 3. Compute expressions for $|e_1\rangle$ and $|e_2\rangle$.
(e) Express the basis vectors $|1\rangle$ and $|2\rangle$ in terms of $|e_1\rangle$ and $|e_2\rangle$.
(f) Express the normalized eigenstates of $B$ in terms of the normalized eigenstates of $A$.
(g) Express $B$ in Dirac notation in terms of $|e_1\rangle$ and $|e_2\rangle$.
(h) Compute the expectation value of $B$ in the state $|e_2\rangle$.

10. Let $L^2(\mathbb{R}, dx)$ denote the Hilbert space of complex valued functions on $\mathbb{R}$ satisfying

$$\int_{\mathbb{R}} |\psi(x)|^2 dx < \infty,$$

for all $\psi(x) \in L^2(\mathbb{R}, dx)$. Consider the operators on $L^2(\mathbb{R}, dx)$ defined by:

$$\mathcal{O}\psi(x) = x^2\psi(x),$$
$$\mathcal{N}\psi(x) = \int_0^x \psi(x')dx',$$

for all $\psi(x) \in L^2(\mathbb{R}, dx)$.

(a) Show that $\mathcal{O}$ is a linear operator.

(b) Show that $\mathcal{N}$ is a linear operator.

(c) Show that $\mathcal{O}$ is self-adjoint.

(d) Does $\mathcal{O}\mathcal{N}$ equal $\mathcal{N}\mathcal{O}$ for all $\psi(x) \in L^2(\mathbb{R}, dx)$? Justify your answer.

11. Consider the complex inner product space $\mathbb{C}^2$ with basis:

$$|e_1\rangle = \frac{1}{\sqrt{2}}\begin{pmatrix} 1 \\ -1 \end{pmatrix}, \quad |e_2\rangle = \frac{1}{\sqrt{2}}\begin{pmatrix} 1 \\ 1 \end{pmatrix}.$$

(a) Using Dirac notation, construct a self-adjoint operator on $\mathbb{C}^2$ with eigenvalues 7 and 3.

(b) Using the explicit row vector and column vector forms for the ket, bra combinations in (a), write the matrix representation of the operator that you constructed in (a).

(c) Using Dirac notation, is it possible to construct a self-adjoint operator on $\mathbb{C}^2$ with eigenvalues 1 and $i$? Justify your answer.

# Chapter 2

# Dynamics of a Quantum Particle

## 2.1 The Schrödinger Equation

The non-relativistic, time dependent Schrödinger equation for a single particle of (constant) mass $m$ in $d$ dimensions (where, practically speaking $d$ will be 1, 2, or 3) has the form:

$$-\frac{\hbar^2}{2m}\nabla^2\psi(\mathbf{r},t) + V(\mathbf{r},t)\psi(\mathbf{r},t) = i\hbar\frac{\partial}{\partial t}\psi(\mathbf{r},t), \qquad (2.1)$$

where $\mathbf{r} \in \mathbb{R}^d$ (or some subset of $\mathbb{R}^d$), $\hbar = \frac{h}{2\pi}$, where $h$ is Planck's constant, and $V(\mathbf{r},t)$ is the potential energy. Schrödinger's original paper on the topic[1] should be mostly understandable to you and provides interesting background leading to the development of the equation (but, in this regard the book by Stone[2] is absolutely superb). Many books offer a description of a type of derivation of the Schrödinger equation. For our purposes, we will just start with the equation, study its structure, solve it in certain situations, and learn how to interpret the results. But concerning the derivation of the Schrödinger equation, the following quote of Richard Feynman[3] is particularly insightful and captures some of the mystery surrounding the development of quantum mechanics, in Feynman's inimitable manner.

---

[1] E. Schrödinger. An undulatory theory of the mechanics of atoms and molecules. *Physical review*, 28(6):1049, 1926.

[2] A D. Stone. *Einstein and the quantum: The quest of the valiant Swabian.* Princeton University Press, 2015.

[3] A. J. G. Hey, T. Hey and P. Walters. *The new quantum universe.* Cambridge University Press, 2003.

> Where did we get that (equation) from? Nowhere. It is
> not possible to derive it from anything you know. It came
> out of the mind of Schrödinger.

The Schrödinger equation is a linear partial differential equation. Boundary conditions and initial conditions are essential for specifying the particular setting for which a solution is sought. These will be dealt with in the particular problems that we study. The solution of (2.1) is a complex valued function, $\psi(\mathbf{r}, t)$, which is referred to as the *wavefunction*. We will discuss the meaning of the wavefunction for a particle of mass $m$ moving under the influence of a potential energy function $V(\mathbf{r}, t)$ shortly. But first we discuss some general features of solving this linear partial differential equation.

The difficulty in solving (2.1) is directly related to the form of the potential energy, $V(\mathbf{r}, t)$. For example, if $V(\mathbf{r}, t) = 0$ (a "free particle") the solution of (2.1) is straightforward. We will treat this situation shortly. However, first we consider the general case where the potential energy is independent of time, i.e.,

$$V(\mathbf{r}, t) = V(\mathbf{r}). \tag{2.2}$$

In this situation the Schrödinger equation can be solved using the method of separation of variables. To apply this method we assume a solution of the form:

$$\psi(\mathbf{r}, t) = \phi(\mathbf{r}) f(t) \tag{2.3}$$

Substituting this into (2.1) gives:

$$f(t) \left( -\frac{\hbar^2}{2m} \nabla^2 + V(\mathbf{r}) \right) \phi(\mathbf{r}) = \phi(\mathbf{r}) \, i\hbar \frac{d}{dt} f(t) \tag{2.4}$$

We divide both sides by $\phi(\mathbf{r}) f(t)$ and obtain:

$$\frac{1}{\phi(\mathbf{r})} \left( -\frac{\hbar^2}{2m} \nabla^2 \phi(\mathbf{r}) + V(\mathbf{r})\phi(\mathbf{r}) \right) = \frac{i\hbar}{f(t)} \frac{df(t)}{dt} \tag{2.5}$$

The left hand side of (2.5) is a function of $\mathbf{r}$ and the right hand side is a function of $t$. Since $\mathbf{r}$ and $t$ are independent variables the two sides of (2.5) must be equal to the same constant, which we call $E$, and we write the resulting two ordinary differential equations separately below:

$$-\frac{\hbar^2}{2m} \nabla^2 \phi(\mathbf{r}) + V(\mathbf{r})\phi(\mathbf{r}) = E\phi(\mathbf{r}). \tag{2.6}$$

$$\frac{d}{dt} f(t) = \frac{E}{i\hbar} f(t) = -\frac{i}{\hbar} E f(t) \Rightarrow f(t) = f(0) e^{-\frac{i}{\hbar} Et}. \tag{2.7}$$

The equation (2.6) is referred to as the *time independent Schrödinger equation*.

Hence a solution of (2.1) is given by:

$$\psi(\mathbf{r}, t) = f(t)\phi(\mathbf{r}) = \phi(\mathbf{r})f(0)e^{-\frac{i}{\hbar}Et}. \tag{2.8}$$

The constant $f(0)$ will henceforth be dropped from our notation since an arbitrary constant can be dealt with through normalization and the choice of initial condition.

States having the form of (2.7) are referred to as *stationary states*. This terminology may appear a curious because, at first, because the word stationary generally means not changing (in time) and (2.7) clearly has a time dependence as a result of the term $e^{-\frac{i}{\hbar}Et}$. The origin of the term *stationary state* comes from the interpretation of the magnitude squared of the wavefunction (2.15) as a probability density described in Section 2.2. Clearly, the magnitude squared of wavefunctions of the form (2.7) is independent of time. It is important to recall why this is the case. It is a direct consequence of the form of the solution of Schrödinger's equation given in (2.7). The form arose from the method that we used for solving Schrödinger's equation, *separation of variables*. This method was possible because the potential energy function was independent of time. Situations where the potential energy is time dependent are important and fascinating, but are beyond the scope of this book. Some background and examples on this topic can be found in the book by David Tannor.[4]

While (2.8) is a solution of (2.1), it is not the most general solution. It is merely a *particular solution*. We explain this statement in a bit more detail. There are two points to consider:

- We have not imposed any boundary and/or initial conditions on the solution.
- The solution (2.8) is given for a particular separation constant $E$.

The first point is dealt with on a problem-by-problem basis. The second point is more fundamental and underlies all of our approaches to "quantum problems".

The mathematical framework for the solution of the time independent Schrödinger equation is an example of a (linear) eigenvalue problem.

---

[4]D. J Tannor. *Introduction to Quantum Mechanics: A Time-dependent Perspective.* University Science Books, 2007.

In particular, we have shown earlier that the mapping:

$$\phi(\mathbf{r}) \mapsto -\frac{\hbar^2}{2m}\nabla^2\phi(\mathbf{r}) + V(\mathbf{r})\phi(\mathbf{r}), \qquad (2.9)$$

is a linear operator on the complex inner product space (i.e. complex Hilbert space) $L^2(D)$, where $D \subset \mathbb{R}^d$ is the spatial domain. Hence, (2.9) has the form of an eigenvalue problem where $\phi(\mathbf{r})$ is the eigenvector with corresponding eigenvalue $E$. We denote the eigenvector-eigenvalue pairs by $(\phi_k(\mathbf{r}), E_k)$. Since (2.9) is a self-adjoint operator (recall, and compare with, Definition 19) it has a complete set of eigenvectors in $L^2(D)$. Therefore every function $\psi(\mathbf{r}, t)$ in $L^2(D)$ can be represented as follows:

$$\psi(\mathbf{r}, t) = \sum_k c_k(t)\phi_k(\mathbf{r}). \qquad (2.10)$$

Substituting this into (2.1) gives:

$$-\frac{\hbar^2}{2m}\sum_k c_k(t)\nabla^2\phi_k(\mathbf{r}) + \sum_k c_k(t)V(\mathbf{r})\phi_k(\mathbf{r}) = i\hbar\sum_k \phi_k(\mathbf{r})\frac{dc_k(t)}{dt}, \quad (2.11)$$

or

$$\sum_k c_k(t)\underbrace{\left(-\frac{\hbar^2}{2m}\nabla^2\phi_k(\mathbf{r}) + V(\mathbf{r})\phi_k(\mathbf{r})\right)}_{E_k\phi_k(\mathbf{r})} = i\hbar\sum_k \phi_k(\mathbf{r})\frac{dc_k(t)}{dt}. \qquad (2.12)$$

Equating the coefficients on the basis vectors on the left and right hand sides gives:

$$i\hbar\frac{dc_k}{dt} = E_k c_k \Rightarrow c_k(t) = c_k(0)e^{-\frac{iE_k}{\hbar}t}. \qquad (2.13)$$

Substituting this into (2.10) gives:

$$\psi(\mathbf{r}, t) = \sum_k c_k(0)e^{-\frac{iE_k}{\hbar}t}\phi_k(\mathbf{r}). \qquad (2.14)$$

### *Physical Requirements for the Mathematical Structure of the Wavefunction*

For the wavefunction to be physically meaningful we will require it to be a continuous and single valued function. We will make the need for this clear when we describe the physical interpretation of the wavefunction in Section 2.2. Furthermore, the wavefunction has additional properties that come from the structure of the partial differential equation that it satisfies.

In particular, if there is a boundary across which the potential energy changes values then the wave function must be continuous across that boundary and remain single valued. Moreover, if the potential energy function is piecewise continuous, it can be shown that all partial derivatives of the wavefunction with respect to the spatial variables must be continuous across the boundary. It is possible to find solutions of the Schrödinger equation that do not satisfy these properties, but they are not considered to be physically meaningful. We will see an example of such a wavefunction shortly.

*For the remainder of this section we restrict ourselves to the one dimensional Schrödinger equation, i.e. $\mathbf{r} \in \mathbb{R}$, and we will refer to the spatial variable as $x$. However, the general concepts and arguments apply in three dimensions.*

## 2.2 The Interpretation of the Wave Function

In quantum mechanics the wavefunction provides the description for the motion of a particle of (constant) mass $m$ in $d$ dimensions, $\psi(\mathbf{r}, t)$, $\mathbf{r} \in \mathbb{R}^d$. Discovering the "recipe" for using the wavefunction to describe the motion of the particle is one of the great achievements of twentieth century science and is discussed in the papers.[5] In contrast to classical mechanics, rather than providing position and velocity as a function of time, quantum mechanics is an inherently statistical theory, and

$$\rho(\mathbf{r}, t) = |\psi(\mathbf{r}, t)|^2. \tag{2.15}$$

is interpreted as the probability density for the position of a particle with wavefunction $\psi(\mathbf{r}, t)$ at time $t$. Note that $|\psi(\mathbf{r}, t)|^2$ is the essential quantity, and not $\psi(\mathbf{r}, t)$ alone. More explicitly, the probability to find the particle in a volume $V \subset \mathbb{R}^d$ at time $t$ is:

$$\int_V |\psi(\mathbf{r}, t)|^2 d^d\mathbf{r}. \tag{2.16}$$

Note that the probabilistic interpretation requires the wavefunction to be normalized, i.e.,

$$\int_{\mathbb{R}^d} |\psi(\mathbf{r}, t)|^2 d^d\mathbf{r} = 1. \tag{2.17}$$

---

[5]Max Born. Quantenmechanik der stoßvorgänge. *Zeitschrift für physik*, 38(11–12): 803–827, 1926.

John Archibald Wheeler and Wojciech Hubert Zurek. *Quantum Theory and Measurement*, volume 40. Princeton University Press, 2014.

for bound systems. Consequently, normalization of wavefunctions will be a constant theme throughout this book. Also, note that because of the absolute value in the integrand of (2.16), multiplying the wavefunction by a factor $e^{i\alpha}$, $\alpha \in \mathbb{R}$, does not change the value of the probability density. Often the phrase, "an overall constant phase is unobservable" is used to describe this feature. This will be another theme that occurs throughout this book.

Once we have a probability density other statistical quantities can be computed and used to describe the motion. For example, the average position of a particle described by a wavefunction $\psi(\mathbf{r}, t)$ at time $t$ is

$$\langle \mathbf{r} \rangle \equiv \mathbf{r}_{\mathrm{av}} = \int_{\mathbb{R}^d} \mathbf{r} |\psi(\mathbf{r}, t)|^2 d^d\mathbf{r}, \tag{2.18}$$

and the variance of the position around this average is:

$$\langle (\mathbf{r} - \langle \mathbf{r} \rangle)^2 \rangle = \langle |\mathbf{r}|^2 \rangle - |\langle \mathbf{r} \rangle|^2$$

$$= \int_{\mathbb{R}^d} |\mathbf{r} - \langle \mathbf{r} \rangle|^2 |\psi(\mathbf{r}, t)|^2 d^d\mathbf{r}. \tag{2.19}$$

## 2.3 The Free Particle

A "free particle" is a particle that moves in a region where it is not subjected to an external force. This means that the potential energy is constant, which we will take to be zero in this discussion. We will treat both the classical and quantum mechanical free particle.

### *Classical Mechanical Free Particle*

Newton's equations of motion for a free particle of (constant) mass $m$ are:

$$m\ddot{\mathbf{r}} = 0, \tag{2.20}$$

or, since $m$ is constant:

$$\frac{d}{dt}(m\dot{\mathbf{r}}) = 0. \tag{2.21}$$

Integrating once gives:

$$m\dot{\mathbf{r}} \equiv \mathbf{p} = \text{constant}, \tag{2.22}$$

where $\mathbf{p}$ is referred to as the (linear) momentum. Hence (2.22) expresses the well-known result that (linear) momentum is constant (in time) for a classical particle of constant mass that is not acted on by a force. Moreover, (2.22) can be integrated to obtain the position of the free particle as a function ot time:

$$\mathbf{r}(t) = \mathbf{r}(0) + \frac{\mathbf{p}}{m}t. \tag{2.23}$$

The total energy of the particle is purely kinetic energy and is given by:

$$E = \frac{1}{2}m\dot{\mathbf{r}} \cdot \dot{\mathbf{r}} \equiv \frac{1}{2}mv^2 = \frac{\mathbf{p} \cdot \mathbf{p}}{2m} \equiv \frac{p^2}{2m} \tag{2.24}$$

where $v$ denotes the magnitude of $\dot{\mathbf{r}}$ and $p$ denotes the magnitude of $\mathbf{p}$.

Hence, we see from (2.22), (2.23), and (2.24) that the momentum, position, and total energy of a classical free particle can be determined simultaneously.

### Quantum Mechanical Free Particle

Now we treat the quantum mechanical free particle of (constant) mass $m$. In this case we will restrict ourselves to the one dimensional case. The three dimensional case is straightforward, but the details tend to distract from the main ideas in the first time that one sees it.

The time independent Schrödinger equation (2.6) in this case is given by:

$$-\frac{\hbar^2}{2m}\frac{d^2}{dx^2}\phi(x) = E\phi(x). \tag{2.25}$$

Initially, there are three cases to consider: $E > 0$, $E < 0$ and the trivial case $E = 0$. It can be shown that the solution for $E < 0$ gives rise to a wavefunction that is non-differentiable at the origin, i.e. it has a cusp. Therefore it is not a physically interesting situation, and we will therefore not consider this case. We will focus on the case $E > 0$, for which we have:

$$\frac{d^2\phi}{dx^2} + k^2\phi = 0, \tag{2.26}$$

where

$$\frac{2mE}{\hbar^2} \equiv k^2 \Rightarrow E = \frac{\hbar^2 k^2}{2m} > 0. \tag{2.27}$$

The solution can be written as:

$$\phi(x) = Ae^{ikx} + Be^{-ikx}, \tag{2.28}$$

where $A$ and $B$ are (complex) constants. Insight into the nature of (2.28) can be obtained by noting that the functions $e^{ikx}$ and $e^{-ikx}$ are eigenfunctions of the momentum operator. This can be seen by a direct calculation. Recall (see Definition 18) that the action of the momentum operator, $P \equiv \frac{\hbar}{i}\frac{d}{dx}$ on a function $\phi(x)$ is of the form:

$$P\phi(x) \equiv \frac{\hbar}{i}\frac{d\phi(x)}{dx}. \tag{2.29}$$

Then we have:

$$\frac{\hbar}{i}\frac{d}{dx}e^{\pm ikx} = \frac{\hbar}{i}\left(\pm ik\right)e^{\pm ikx} = \pm\hbar k e^{\pm ikx}. \tag{2.30}$$

Hence $e^{ikx}$ is an eigenfunction with eigenvalue $\hbar k$. This corresponds to a particle moving to the right with positive momentum $\hbar k$. Similarly, $e^{-ikx}$ is an eigenfunction with eigenvalue $-\hbar k$, corresponding to a particle moving to the left with negative momentum $-\hbar k$.

The quantum mechanical free particle moving to the right with momentum $\hbar k$ is described by the wavefunction $\psi(x) = Ae^{ikx}$. The momentum and the energy, (2.27), are known precisely, and the average position can be determined from the probability density:

$$|\psi(x)|^2 = |A|^2. \tag{2.31}$$

This expression raises a number of issues.[6] The probability density is constant. Hence the probability of finding a particle in any interval on the $x$ axis is the same as that for any other interval of equal length, and does not change with time. In other words, there are no special places for a free particle to be found. However, the concept of "probabilty" is questionable in this case since the wavefunction is not normalizable, i.e. the integral of the probability over the entire line is infinite. So while the momentum and energy can be known precisely, the position is unknown.

So how can we describe a quantum mechanical free particle in a way that bears some resemblance to our description of a classical free particle? This can be done using the notion of a *wave packet*.

### *Wave Packets*

A wave packet is constructed by considering a sum of functions of the form $a(k)e^{ikx}$, where the sum is over $k$ (hence, particle momenta). Because $k$ is

---

[6]Despite the fact that functions of the form $e^{\pm ikx}$ are not normalizable on their entire domain they still play an important role in quantum mechanics. We will see this shortly when we introduce the concept of a wave packet. However, there has been significant effort in providing mathematical meaning for the normalization of such "non-normalizable" functions. This is described in detail in Ref. 1.

unrestricted, the sum actually is an integral and we write

$$\psi(x,0) = \int_{-\infty}^{\infty} a(k)e^{ikx}\,dk. \tag{2.32}$$

The amplitudes $a(k)$ of each $e^{ikx}$ determine the so-called spectral content of the wave packet. We will assume $a(k)$ is a Gaussian function of the following form:

$$a(k) = \frac{C\alpha}{\sqrt{\pi}}e^{-\alpha^2 k^2}, \tag{2.33}$$

where $C$ and $\alpha$ are real constants that can be used to control the shape of the wave packet. The constant $C$ is referred to as the amplitude. Note that at $x = \pm 2\alpha$ the amplitude is reduced by a factor $\frac{1}{e}$. This leads us to identify $\alpha$ with the width of the Gaussian.

With the choice of $a(k)$ to be a Gaussian function, the phrase *Gaussian wavepacket* is often used. Substituting into gives:

$$\psi(x,0) = \int_{-\infty}^{\infty} a(k)e^{ikx}\,dk = \frac{C\alpha}{\sqrt{\pi}} \int_{-\infty}^{\infty} e^{(ikx-\alpha^2 k^2)}\,dk. \tag{2.34}$$

To evaluate the integral, we first complete the square in the exponent as follows:

$$ikx - \alpha^2 k^2 = -\left(\alpha k - \frac{ix}{2\alpha}\right)^2 - \frac{x^2}{4\alpha^2}. \tag{2.35}$$

The second term on the right is constant for the integration over $k$. To integrate the first term we change variables with the substitution $z = \alpha k - \frac{ix}{2\alpha}$, we obtain

$$\psi(x,0) = \frac{C}{\sqrt{\pi}}e^{-\frac{x^2}{4\alpha^2}} \int_{-\infty}^{\infty} e^{-z^2}\,dz. \tag{2.36}$$

The integral is well-known[7] and is equal to $\sqrt{\pi}$, which yields

$$\psi(x,0) = Ce^{-\frac{x^2}{4\alpha^2}} = Ce^{-\left(\frac{x}{2\alpha}\right)^2}, \tag{2.37}$$

and the constant $C$ can be chosen so that the wavefunction is normalized. In other words, we compute:

$$\int_{-\infty}^{\infty} (\psi(x,0))^2\,dx = C^2 \int_{-\infty}^{\infty} e^{-2\frac{x^2}{4\alpha^2}}\,dx = 1 \tag{2.38}$$

---

[7] $\int_{-\infty}^{\infty} e^{-x^2}\,dx = \sqrt{\pi}$ is the famous Gaussian integral. It is famous because it turns up over and over again in (seemingly) diverse areas of mathematics and physics.

Using the expression for the Gaussian integral, we find that

$$C = \frac{1}{\sqrt{\alpha}\,(2\pi)^{\frac{1}{4}}},$$

and, hence,

$$\psi(x,0) = \frac{1}{\sqrt{\alpha}\,(2\pi)^{\frac{1}{4}}} e^{-\left(\frac{x}{2\alpha}\right)^2}. \tag{2.39}$$

Wave packets are the basic building blocks of a number of useful tools for understanding the time evolution of quantum systems. More background, and many examples, can be found in the books of Klauder and Skagerstam[8] and Tannor[9] and the comprehensive review paper of Gilmore and Zhang.[10] We will learn more about the role played by the wave packet description of particles in determining position and momentum when we consider the issue of uncertainty in Section 3.3. This will motivate the fascinating phenomenon that propagation of wavepackets in time has a remarkable connection to the propagation of classical particles according to classical mechanics. This allows for a certain classical intuition for quantum phenomena. In addition to the references given above, the following series of papers by Nieto and collaborators are insightful.[11]

Finally, Heller and collaborators[12] have shown how Gaussian wave packet propagation can provide remarkable insights into important quantum dynamics problems in chemical dynamics.

---

[8] John R Klauder and Bo-Sture Skagerstam. *Coherent states: applications in physics and mathematical physics.* World scientific, 1985.

[9] D. J Tannor. *Introduction to Quantum Mechanics: A Time-dependent Perspective.* University Science Books, 2007.

[10] Wei-Min Zhang and Robert Gilmore. Coherent states: Theory and some applications. *Reviews of Modern Physics*, 62(4):867, 1990.

[11] Michael Martin Nieto and L.M. Simmons Jr. Coherent states for general potentials. I. formalism. *Physical Review D*, 20(6):1321, 1979b.

Michael Martin Nieto and L.M. Simmons Jr. Coherent states for general potentials. II. confining one-dimensional examples. *Physical Review D*, 20(6):1332, 1979c.

Michael Martin Nieto and L. M. Simmons Jr. Coherent states for general potentials. III. nonconfining one-dimensional examples. *Physical Review D*, 20(6):1342, 1979a.

Michael Martin Nieto. Coherent states for general potentials. IV. three-dimensional systems. *Physical Review D*, 22(2):391, 1980.

Vincent P Gutschick and Michael Martin Nieto. Coherent states for general potentials. v. time evolution. *Physical Review D*, 22(2):403, 1980.

[12] Eric J. Heller. Time-dependent approach to semiclassical dynamics. *The Journal of Chemical Physics*, 62(4):1544–1555, 1975.

Eric J. Heller. Time dependent variational approach to semiclassical dynamics. *The Journal of Chemical Physics*, 64(1):63–73, 1976.

## 2.4 The Square Well

The square well potential is treated in every elementary quantum mechanics textbook that we are aware of. Quoting from,[13]

> "This potential is awfully artificial, but I urge you to treat it with respect. Despite its simplicity — or rather, precisely because of its simplicity — it serves as a wonderfully accessible test case for all the fancy stuff that comes later."

We will see this very explicitly in this chapter since we will use the square well to develop much of the mathematical structure of the Schrödinger equation. The work[14] is a real tour de force on the square well and provides a discussion of interesting physical situations and applications where the square well, or related models, provides physical and mathematical insight. Notably, this paper points to the origin of the square well in work of Mott,[15] a Nobel Laureate and former head of the Wills Physical Laboratory at the University of Bristol. We now turn to the description of a quantum particle moving in an infinite square well.

We consider a particle moving in the interval $[0, a]$, where $V(x) = 0$ in this interval, and $V(x) = \infty$ for $x > a$ and $x < 0$. The time independent Schrödinger equation inside the well is:

$$-\frac{\hbar^2}{2m}\frac{d^2}{dx^2}\psi = E\psi, \tag{2.40}$$

where we will take as boundary conditions $\psi(a) = \psi(0) = 0$. The general solution of (2.40) is:

$$\psi(x) = \begin{cases} A\cosh\left(\sqrt{2m|E|}\,\frac{x}{\hbar}\right) + B\sinh\left(\sqrt{2m|E|}\,\frac{x}{\hbar}\right) & \text{if } E < 0 \\ A + Bx & \text{if } E = 0 \\ A\cos\left(\sqrt{2mE}\,\frac{x}{\hbar}\right) + B\sin\left(\sqrt{2mE}\,\frac{x}{\hbar}\right) & \text{if } E > 0 \end{cases} \tag{2.41}$$

---

Daniel Huber, Eric J. Heller and Robert G. Littlejohn. Generalized Gaussian wave packet dynamics, schrödinger equation, and stationary phase approximation. *The Journal of chemical physics*, 89(4):2003–2014, 1988.

E.J. Heller. Guided Gaussian wave packets. *Accounts of chemical research*, 39(2): 127–134, 2006.

[13]D. J. Griffiths and D. F. Schroeter. *Introduction to quantum mechanics*. Cambridge University Press, 2018.

[14]M. Belloni and R. W. Robinett. The infinite well and Dirac delta function potentials as pedagogical, mathematical and physical models in quantum mechanics. *Physics Reports*, 540(2):25–122, 2014.

[15]Nevill Francis Mott. *An Outline of Wave Mechanics*. The University Press, 1930.

when

$$E \le 0 \quad \psi(0) = \psi(a) = 0 \quad \Rightarrow A = B = 0, \tag{2.42}$$

and when

$$E > 0 \quad \begin{aligned} \psi(0) &= 0 \Rightarrow A = 0, \\ \psi(a) &= 0 \Rightarrow \sqrt{2mE}\,\tfrac{a}{\hbar} = n\pi. \end{aligned} \tag{2.43}$$

This implies that we have nontrivial solutions if and only if[16]:

$$E = E_n = \frac{n^2\pi^2\hbar^2}{2ma^2}, \quad n = 1, 2, \ldots, \tag{2.44}$$

and therefore

$$\psi_n(x) = B_n \sin\left(\frac{n\pi x}{a}\right). \tag{2.45}$$

The constant $B_n$ is determined by requiring

$$\int_0^a |\psi_n(x)|^2 dx = 1. \tag{2.46}$$

which gives:

$$\frac{1}{2}|B_n|^2 a = 1, \tag{2.47}$$

or

$$B_n = \sqrt{\frac{2}{a}}, \tag{2.48}$$

and therefore

$$\psi_n(x) = \sqrt{\frac{2}{a}} \sin\left(\frac{n\pi x}{a}\right). \tag{2.49}$$

It is an easy calculation to verify that the $\psi_n(x)$ form an orthonormal set of functions, i.e.,

$$\int_0^a \overline{\psi_n(x)}\psi_m(x)dx = \delta_{n,m}. \tag{2.50}$$

This property will be central to many calculations that we carry out for the square well.

---

[16]The "energy eigenvalue" expression for the square well, (2.44), highlights the quantum aspect of the square well. The energy only exists in discrete amounts. Note that the energy cannot be zero. The integer $n$ is often referred to as the *principal quantum number*.

Using (2.8), the time evolution of $\psi_n(x)$ is given by:

$$\psi_n(x,t) = \sqrt{\frac{2}{a}} \sin\left(\frac{n\pi x}{a}\right) e^{-\frac{in^2\pi^2\hbar t}{2ma^2}}.$$ (2.51)

Linearity of the Schrödinger equation implies that the general wavefunction can be obtained by a superposition of the eigenfunctions:

$$\psi(x,t) = \sqrt{\frac{2}{a}} \sum_{n=1}^{\infty} c_n e^{-\frac{in^2\pi^2\hbar t}{2ma^2}} \sin\left(\frac{n\pi x}{a}\right).$$ (2.52)

The expansion coefficients $c_n$ can be obtained as follows. Noting that

$$\psi(x,0) = \sqrt{\frac{2}{a}} \sum_{n=1}^{\infty} c_n \sin\left(\frac{n\pi x}{a}\right),$$ (2.53)

therefore

$$c_n = \sqrt{\frac{2}{a}} \int_0^a \psi(x,0) \sin\left(\frac{n\pi x}{a}\right) dx.$$ (2.54)

Summarizing: for the square well the time-independent Schrödinger equation has eigenvalues $E_n = \frac{n^2\pi^2\hbar^2}{2ma^2}$, $n = 1, 2, \ldots$ with corresponding eigenfunctions $\psi_n(x) = \sqrt{\frac{2}{a}} \sin\left(\frac{n\pi x}{a}\right)$. The general solution of the time-dependent Schrödinger equation can be written as $\psi(x,t) = \sum_n c_n e^{\frac{-iE_n t}{\hbar}} \psi_n(x)$ with $c_n = \int_0^a \psi(x,0)\psi_n(x)dx$.

### Interpretation of the Wavefunction for the Square Well

Now we apply the general probabilistic interpretation of the wavefunction described in Section 2.2 to the quantum particle in the square well.

Using (2.15) and (2.51), the probability density for a particle in the $n^{\text{th}}$ eigenstate of a square well is:

$$|\psi_n(x)|^2 = \frac{2}{a} \sin^2\frac{n\pi x}{a} = \frac{1}{a}\left(1 - \cos\frac{2n\pi x}{a}\right).$$ (2.55)

1. We can integrate (2.55) to obtain the probability distribution function, which gives the probability of finding a particle between $x = 0$ and $x = X \leq a$:

$$\int_0^X |\psi_n(x)|^2 dx = \frac{X}{a} - \frac{1}{2n\pi} \sin\frac{2n\pi X}{a} \equiv P_n(X).$$ (2.56)

2. The mean position of the particle is:

$$\int_0^a x|\psi_n(x)|^2 dx = \frac{1}{a}\int_0^a \left(x - x\cos\frac{2n\pi x}{a}\right)dx = \frac{a}{2}. \qquad (2.57)$$

Note that it is independent of $n$.

3. The variance of the position is:

$$\int_0^a \left(x - \frac{1}{2}a\right)^2 |\psi_n(x)|^2 dx = \left(\frac{1}{12} - \frac{1}{2n^2\pi^2}\right)a^2, \qquad (2.58)$$

which does depend on $n$.

Now note that the general time-dependent solution of the Schrödinger equation for the square well is given by:

$$\psi(x,t) = \sum_n c_n \psi_n(x)\exp\left(-i\frac{E_n t}{\hbar}\right)$$

$$= \sqrt{\frac{2}{a}}\sum_n c_n \sin\frac{n\pi x}{a}\exp\left(-i\frac{n^2\pi^2\hbar t}{2ma^2}\right). \qquad (2.59)$$

Then, using orthogonality of the $\psi_n(x)$, the probability of finding a particle at some location in the square well is given by:

$$\int_0^a |\psi(x,t)|^2 dx = \int_0^a \sum_{n,m} c_n \bar{c}_m \psi_n(x)\overline{\psi_m(x)}\exp\left(-i\frac{(E_n - E_m)t}{\hbar}\right)dx$$

$$= \sum_m |c_m|^2. \qquad (2.60)$$

If we require

$$\int_0^a |\psi(x,t)|^2 dx = 1, \qquad (2.61)$$

then this implies that

$$\sum_m |c_m|^2 = 1. \qquad (2.62)$$

This leads to the following interpretation.[17] The probability of measuring the energy to be $E_n$ or, equivalently, of finding the particle to be in the $n^{\text{th}}$ eigenstate, $\psi_n(x)$, is $|c_n|^2$.

---

[17]This is a very important point. It says that the probability of measuring the value $E_n$ is $|c_n|^2$. This interpretation is due to Heisenberg and Born.

## 2.5 Some Additional Properties of the Square Well Directly from the Structure of the Schrödinger Equation

We derived the following properties of the square well directly from the solutions of the Schrödinger equation for the square well:

1. The energy levels are non-degenerate, i.e., $E_n \neq E_m$ if $n \neq m$.
2. The energy levels are real, i.e. $E_n \in \mathbb{R}$, $n = 1, 2, \ldots$.
3. The eigenfunctions are orthonormal, i.e.,

$$\int_0^a \overline{\psi_n(x)} \psi_m(x)\, dx = \delta_{n,m}. \tag{2.63}$$

We now want to show that these properties follow directly from the structure of the Schrödinger equation.

**Theorem 2.** *The energy levels of the one dimensional square well are non-degenerate.*

**Proof.** We argue by contradiction.

Assume that eigenfunctions $\psi_m(x)$ and $\psi_n(x)$ corresponding to different states $(n \neq m)$ have $E_n = E_m$:

$$-\frac{d^2\psi_n}{dx^2} = \frac{2mE_n}{\hbar^2}\psi_n,$$

$$-\frac{d^2\psi_m}{dx^2} = \frac{2mE_m}{\hbar^2}\psi_m, \qquad E_n = E_m. \tag{2.64}$$

Multiplying the first equation in (2.64) by $\psi_m(x)$ and the second equation in (2.64) by $\psi_n(x)$ and subtracting gives:

$$\psi_m \frac{d^2\psi_n}{dx^2} - \psi_n \frac{d^2\psi_m}{dx^2} = 0, \tag{2.65}$$

or

$$\frac{d}{dx}\left(\psi_m \frac{d\psi_n}{dx} - \psi_n \frac{d\psi_m}{dx}\right) = 0, \tag{2.66}$$

from which it follows that:

$$\psi_m \frac{d\psi_n}{dx} - \psi_n \frac{d\psi_m}{dx} = \text{constant}. \tag{2.67}$$

If we evaluate (2.67) at $x = 0$ where $\psi_m(0) = \psi_n(0) = 0$ we conclude that constant $= 0$. Therefore (2.67) becomes:

$$\psi_m \frac{d\psi_n}{dx} - \psi_n \frac{d\psi_m}{dx} = 0. \tag{2.68}$$

Dividing (2.68) by $\psi_n(x)\psi_m(x)$ gives:

$$\frac{1}{\psi_n} \frac{d\psi_n}{dx} = \frac{1}{\psi_m} \frac{d\psi_m}{dx}. \tag{2.69}$$

After integrating this equation we obtain:

$$\log|\psi_n| = \log|\psi_m| + \text{constant}, \tag{2.70}$$

or

$$\psi_n(x) = \text{constant } \psi_m(x). \tag{2.71}$$

from which it follows that $\psi_n(x)$ and $\psi_m(x)$ represent the same state, since they are equal *after normalization*. This contradicts our original assumption. $\qquad\square$

The following theorem is concerned with non-degeneracy of the energy levels of a more general one dimensional potential energy function, $V(x)$, that has the property that the Schrödinger equation with this potential has a set of eigenfunctions and eigenvalues, $(\phi_k(x), E_k)$.

**Theorem 3.** *The energy levels of the one dimensional time independent Schrödinger equation:*

$$-\frac{\hbar^2}{2m} \frac{d^2\psi}{dx^2} + V(x)\psi(x) = E\psi(x), \tag{2.72}$$

*with boundary condition $\psi(x) \to 0$ as $|x| \to \infty$ are all non-degenerate.*

**Proof.** We will leave the proof of this result as an exercise. $\qquad\square$

Non-degeneracy of the energy levels is a general result in one dimension, but it is not necessarily true in higher dimensions.

Next we will prove a basic result that will enable us to prove several properties of the eigenstates of the square well.

First, let $\alpha(x)$ and $\beta(x)$ be any two functions defined on $[0, a]$ which vanish at $x = 0$ and $x = a$ and which are twice differentiable. Then we have the following theorem.

**Theorem 4 (Hermiticity).** *The differential operator:*

$$H = -\frac{\hbar^2}{2m}\frac{d^2}{dx^2}, \tag{2.73}$$

*satisfies*

$$\int_0^a \overline{\alpha} H \beta \, dx = \int_0^a \overline{(H\alpha)}\beta \, dx, \tag{2.74}$$

*where the overline indicates the complex conjugate.*[18]

The property embodied by (2.74) is often referred to as *Hermiticity.*[19]

**Proof.** We use integration by parts:

$$-\frac{\hbar^2}{2m}\int_0^a \overline{\alpha}\frac{d^2\beta}{dx^2}dx = -\frac{\hbar^2}{2m}\left\{\left[\overline{\alpha}\frac{d\beta}{dx}\right]_0^a - \int_0^a \frac{d\overline{\alpha}}{dx}\frac{d\beta}{dx}dx,\right\}, \tag{2.75}$$

where the first term in the curly brackets is zero since $\overline{\alpha(a)} = \overline{\alpha(0)} = 0$. We then integrate the remaining terms by parts:

$$-\frac{\hbar^2}{2m}\left\{-\left[\frac{d\overline{\alpha}}{dx}\beta\right]_0^a + \int_0^a \frac{d^2\overline{\alpha}}{dx^2}\beta \, dx\right\}, \tag{2.76}$$

where, again, the first term in the curly brackets is zero since $\overline{\beta(a)} = \overline{\beta(0)} = 0$. Hence, we finally obtain:

$$= -\frac{\hbar^2}{2m}\int_0^a \overline{\left(\frac{d^2\alpha}{dx^2}\right)}\beta \, dx, \tag{2.77}$$

which was the result to be proved. $\qquad\square$

Now we describe some further consequences of this theorem for the square well.

1. The energy levels of the square well are all real. This can be seen by a direct calculation and the use of Theorem 4. We have:

$$H\psi_n = E_n\psi_n \quad \text{and} \quad \int_0^a \overline{\psi}_n\psi_n \, dx = 1, \tag{2.78}$$

---

[18] Note that (2.73) is the Hamiltonian for the square well.
[19] You should compare (2.74) with the general definition of a self-adjoint or Hermitian operator given in Definition 15.

from which it follows that:

$$E_n = \int_0^a \bar{\psi}_n H \psi_n dx. \tag{2.79}$$

Now we apply Theorem 4 to this result with $\alpha = \beta = \psi_n$ (which satisfies the hypotheses of the theorem) to obtain:

$$E_n = \int_0^a \left(\overline{H\psi_n}\right) \psi_n dx = \int_0^a \bar{E}_n \bar{\psi}_n \psi_n dx = \bar{E}_n, \tag{2.80}$$

from which it follows that the energy levels are real.[20]

2. Now we show that the eigenstates satisfy an orthonormality condition, i.e.,

$$\int_0^a \bar{\psi}_m \psi_n dx = \delta_{mn}. \tag{2.81}$$

We can assume that we have normalized the eigenstates and that we have

$$\int_0^a \bar{\psi}_n \psi_n dx = 1, \tag{2.82}$$

therefore we only need to consider the case $n \neq m$. Integrating the expression $H\psi_n = E_n \psi_n$ gives:

$$E_n \int_0^a \bar{\psi}_m \psi_n dx = \int_0^a \bar{\psi}_m H \psi_n dx. \tag{2.83}$$

Applying Theorem 4 to this last expression, and using the fact that $\overline{E_n} = E_n$ gives

$$\int_0^a \left(\overline{H\psi_m}\right) \psi_n dx = E_m \int_0^a \bar{\psi}_m \psi_n dx \tag{2.84}$$

Hence

$$(E_n - E_m) \int_0^a \bar{\psi}_m \psi_n dx = 0. \tag{2.85}$$

Since the energy levels are non-degenerate we have[21]:

$$\int_0^a \bar{\psi}_m \psi_n dx = 0 \quad \text{if} \quad m \neq n. \tag{2.86}$$

---

[20] You should compare this calculation with 1.63.
[21] You should compare this calculation with 1.65.

This result for the square well can be generalized as follows. Let $H = -\frac{\hbar^2}{2m}\frac{d^2}{dx^2} + V(x)$, and $\alpha(x)$ and $\beta(x)$ be any two functions, defined on $\mathbb{R}$, that are twice differentiable and satisfy $\alpha(x) \to 0$, $\beta(x) \to 0$ as $|x| \to \infty$. Then we can prove the following result

**Theorem 5.**

$$\int_{-\infty}^{+\infty} \bar{\alpha} H \beta \, dx = \int_{-\infty}^{+\infty} \left(\overline{H\alpha}\right) \beta \, dx \tag{2.87}$$

**Proof.** We will leave the proof of this result as an exercise. $\qquad\square$

Theorem 5 has consequences similar to the consequences of Theorem 4 for the square well. In particular, The energy levels $E_n$ and the corresponding eigenfunctions $\psi_n$ satisfying the time independent Schrödinger equation

$$H\psi_n = E_n\psi_n, \tag{2.88}$$

have the following properties:

1. $E_n \in \mathbb{R}$.
2. $\int_{-\infty}^{+\infty} \overline{\psi_m(x)}\psi_n(x)dx = \delta_{m,n}$.

## 2.6 Probability Current and the Conservation of Probability

We have seen that the probability density, (2.15), plays a central role in how we relate the wavefunction to observable properties of a quantum particle. In this section we explore properties of the time evolution of the probability density.

The first question we consider is the following. If

$$\int_{\mathbb{R}^d} |\psi(\mathbf{r},0)|^2 d^d\mathbf{r} = 1. \tag{2.89}$$

then under what conditions do we have

$$\int_{\mathbb{R}^d} |\psi(\mathbf{r},t)|^2 d^d\mathbf{r} = 1, \qquad \forall\, t > 0? \tag{2.90}$$

If the wavefunction from which the probability density is constructed is an eigenstate (where the potential energy is independent of time) then the answer is clear

$$|\psi_n(\mathbf{r},t)|^2 = |\psi_n(\mathbf{r})e^{-\frac{iE_n t}{\hbar}}|^2 = |\psi_n(\mathbf{r})|^2, \tag{2.91}$$

i.e., the probability density is independent of time (but it depends on the spatial variable).

Now we will derive a partial differential equation that describes the evolution of the probability density. We begin by computing the time derivative of $\rho$:

$$\frac{\partial}{\partial t}\rho = \frac{\partial}{\partial t}(\psi\bar{\psi}) = \frac{\partial\psi}{\partial t}\bar{\psi} + \psi\frac{\partial\bar{\psi}}{\partial t}. \tag{2.92}$$

Recall from (2.1) that the Schrödinger equation and its complex conjugate are given by:

$$\frac{\partial\psi}{\partial t} = \frac{1}{i\hbar}\left(-\frac{\hbar^2}{2m}\nabla^2\psi + V\psi\right), \tag{2.93}$$

and

$$\frac{\partial\bar{\psi}}{\partial t} = -\frac{1}{i\hbar}\left(-\frac{\hbar^2}{2m}\nabla^2\bar{\psi} + V\bar{\psi}\right). \tag{2.94}$$

Substituting these two equations into (2.92) gives:

$$\frac{\partial\rho}{\partial t} = -\frac{\hbar}{2mi}\left(\bar{\psi}\nabla^2\psi - \psi\nabla^2\bar{\psi}\right),$$

$$= -\frac{\hbar}{2mi}\nabla\cdot\left(\bar{\psi}\nabla\psi - \psi\nabla\bar{\psi}\right). \tag{2.95}$$

This leads to the following definition:

**Definition 21 (Probability Current).** The probability current, $\mathbf{j}(\mathbf{r}, t)$, is defined as

$$\mathbf{j}(\mathbf{r}, t) = \frac{\hbar}{2mi}\left(\bar{\psi}\nabla\psi - \psi\nabla\bar{\psi}\right) = \frac{1}{m}\mathrm{Re}\left(\bar{\psi}\frac{\hbar}{i}\nabla\psi\right). \tag{2.96}$$

The calculations above can be collected together into the following theorem.

**Theorem 6.** $\rho(\mathbf{r}, t)$ *and* $\mathbf{j}(\mathbf{r}, t)$ *satisfy*

$$\frac{\partial\rho}{\partial t} + \nabla\cdot\mathbf{j}(\mathbf{r}, t) = 0. \tag{2.97}$$

(2.97) has the same form as the continuity equation from fluid mechanics, which expresses the conservation of mass density under a flow. In this setting the role of mass density is played by the probability density and the role of the flow is played by the probability current. Next we describe how the partial differential equation implies the conservation of the probability.

First we establish some notation. Let $B(R)$ denote the ball of radius $R$ in $\mathbb{R}^d$ and let $S \equiv \partial B(R)$ denote the surface of this ball. Our result will require an assumption on the class of functions under consideration. We will assume the following behavior at infinity for the probability current:

$$\mathbf{j}(\mathbf{r},t)|\mathbf{r}|^{d-1} \to 0 \quad \text{as} \quad |\mathbf{r}| \to \infty. \tag{2.98}$$

Integrating (2.97) over $B(R)$ gives:

$$\frac{\partial}{\partial t} \int_{B(R)} \rho(\mathbf{r},t) d^d\mathbf{r} = - \int_{B(R)} \nabla \cdot \mathbf{j}(\mathbf{r},t) d^d\mathbf{r} = - \int_{\partial B(R)} \mathbf{j}(\mathbf{r},t) \cdot d\mathbf{S}, \tag{2.99}$$

where the last equality follows from the application of the divergence theorem. Now taking $R \to \infty$ and using (2.98) gives:

$$\frac{\partial}{\partial t} \int_{\mathbb{R}^d} \rho(\mathbf{r},t) d^d\mathbf{r} = 0, \tag{2.100}$$

Therefore

$$\int_{\mathbb{R}^d} \rho(\mathbf{r},t) d^d\mathbf{r}, \tag{2.101}$$

is independent of $t$. In particular, if it is 1 for $t = 0$ then it is 1 for all $t$.

The probability current has a suggestive interpretation in terms of classical mechanics. Rewriting the probability current here for easy reference:

$$\mathbf{j}(\mathbf{r},t) = \frac{1}{m} \text{Re} \left( \bar{\psi} \left( \frac{\hbar}{i} \nabla \psi \right) \right). \tag{2.102}$$

Recall that $\frac{\hbar}{i} \nabla \psi$ is interpreted as the product of the momentum with the wavefunction. Therefore $\mathbf{j}(\mathbf{r},t)$ has the form of $\frac{\text{momentum}}{m} |\psi|^2$, i.e. the product of the velocity and the probability density.

## 2.7 Expectation Values

Recall that the average position of a particle in the square well at time $t$ in the state $\psi(x,t)$ is given by:

$$\langle x \rangle = \int_0^a x |\psi(x,t)|^2 dx = \int_0^a \overline{\psi(x,t)} x \psi(x,t) dx. \tag{2.103}$$

$\langle x \rangle$ is called the expectation value of $x$, and it is understood that this requires specifying a specific state, $\psi(x,t)$, for its computation.

**Definition 22 (Expectation Value of Momentum).** We define the expectation value of the momentum $p$ (i.e. the average value of the momentum) at time $t$ as

$$\langle p \rangle = m \frac{d\langle x \rangle}{dt}. \tag{2.104}$$

We have the following characterization of the expectation value of the momentum of a particle in the square well.

**Theorem 7.** *For a particle in a square well described by a normalized wave function $\psi(x,t)$, the expectation value of $p$ in the state $\psi(x,t)$ is given by:*

$$\langle p \rangle = \int_0^a \overline{\psi} \left( -i\hbar \frac{\partial \psi}{\partial x} \right) dx. \tag{2.105}$$

**Proof.** Differentiating (2.103) with respect to time gives:

$$\frac{d\langle x \rangle}{dt} = \int_0^a x \frac{\partial \overline{\psi}}{\partial t} \psi dx + \int_0^a x \overline{\psi} \frac{\partial \psi}{\partial t} dx \tag{2.106}$$

Recall that the Schrödinger equation and its complex conjugate are given by the following two equations:

$$\frac{\partial \psi}{\partial t} = \frac{1}{i\hbar} \left( -\frac{\hbar^2}{2m} \frac{\partial^2 \psi}{\partial x^2} + V\psi \right) \tag{2.107}$$

$$\frac{\partial \overline{\psi}}{\partial t} = -\frac{1}{i\hbar} \left( -\frac{\hbar^2}{2m} \frac{\partial^2 \overline{\psi}}{\partial x^2} + V\overline{\psi} \right) \tag{2.108}$$

Substituting these two equations into (2.106) and multiplying by $m$ gives (note that $V = 0$ for the square well):

$$\langle p \rangle = \frac{\hbar}{2i} \int_0^a x\psi \frac{\partial^2 \overline{\psi}}{\partial x^2} dx - \frac{\hbar}{2i} \int_0^a x\overline{\psi} \frac{\partial^2 \psi}{\partial x^2} dx \tag{2.109}$$

Using Theorem 4 for the first integral gives:

$$\langle p \rangle = \frac{\hbar}{2i} \int_0^a \overline{\psi} \left( \frac{\partial^2}{\partial x^2}(x\psi) - x \frac{\partial^2 \psi}{\partial x^2} \right) dx,$$

$$= \frac{\hbar}{2i} \int_0^a \overline{\psi} \left( x \frac{\partial^2 \psi}{\partial x^2} + 2 \frac{\partial \psi}{\partial x} - x \frac{\partial^2 \psi}{\partial x^2} \right) dx,$$

$$= \frac{\hbar}{i} \int_0^a \overline{\psi} \frac{\partial \psi}{\partial x} dx. \tag{2.110}$$

$\square$

We make the following remarks.

1. The proof uses only the Schrödinger equation and Theorem 4, and it generalizes to higher dimensions in a straightforward manner.
2. The generalization to higher dimensions is

$$\langle \mathbf{p} \rangle = \int_{\mathbb{R}^d} \overline{\psi} \left( \frac{\hbar}{i} \nabla \psi \right) d^d \mathbf{r}. \tag{2.111}$$

3. Note that this result is consistent with the observation that $\frac{\hbar}{i} \nabla \psi =$ momentum $\times \psi$.

This suggests that the quantum expectation value of the energy of a particle in a square well in a state described by $\psi(x,t)$ is:

$$\langle E \rangle = \int_0^a \overline{\psi} \left( -\frac{\hbar^2}{2m} \frac{\partial^2}{\partial x^2} \psi \right) dx = \int_0^a \overline{\psi} i\hbar \frac{\partial \psi}{\partial t} dx. \tag{2.112}$$

Note that is we expand $\psi(x,t)$ in terms of eigenfunctions $\psi_n(x)$

$$\psi(x,t) = \sum_n c_n \psi_n(x) e^{-i\frac{E_n t}{\hbar}}. \tag{2.113}$$

we have, using $\int_0^a \overline{\psi_n(x)} \psi_m(x) dx = \delta_{nm}$

$$\langle E \rangle = \sum_n |c_n|^2 E_n. \tag{2.114}$$

which is consistent with the assumption that $|c_n|^2$ is the probability of measuring the particle to be in the $n^{\text{th}}$ eigenstate.

The generalization to a particle described by the classical Hamiltonian

$$H(q,p) = \frac{p^2}{2m} + V(q), \tag{2.115}$$

is given by:

$$\langle E \rangle = \int_{\mathbb{R}^d} \overline{\psi} \left( -\frac{\hbar^2}{2m} \nabla^2 + V(\mathbf{q}) \right) \psi d^d \mathbf{q} = \int_{\mathbb{R}^d} \overline{\psi} \left( i\hbar \frac{\partial \psi}{\partial t} \right) d^d \mathbf{q}. \tag{2.116}$$

## 2.8  Scattering and Tunneling

In this section several piecewise constant potential energy functions are considered that allow us to describe and analyze the phenomena of scattering and tunneling. Towards this end, the notion of the probability current will play an important role.

However, we first begin with an important observation related to classical mechanical motion. Consider one dimensional motion defined by the Hamiltonian function:

$$H = \frac{p^2}{2m} + V(x), \quad (x, p) \in \mathbb{R}^2. \tag{2.117}$$

For a fixed energy, $E$, i.e., $\frac{p^2}{2m} + V(x) = E$, classical motion is allowable for $E \geq V$ since

$$p = \pm\sqrt{2m(E - V)}. \tag{2.118}$$

In Quantum Mechanics, this constraint can be violated, as we will see.

### Constant Potential

We begin by considering a simple case that will allow us to establish some of the basic ideas that we will use in more complicated situations afterwards. We consider $V(x) = V = $ constant, for which the time independent Schrödinger equation is given by

$$-\frac{\hbar^2}{2m}\frac{d^2\psi}{dx^2} + V\psi = E\psi. \tag{2.119}$$

We consider the two cases.

$\boxed{E > V.}$

We define the wavenumber as $k = \frac{1}{\hbar}\sqrt{2m(E - V)}$. In this case the time independent Schrödinger equation takes the form:

$$-\frac{d^2\psi}{dx^2} = k^2\psi, \tag{2.120}$$

which has the solution

$$\psi(x) = Ae^{ikx} + Be^{-ikx}. \tag{2.121}$$

As discussed in Section 2.3, it follows from

$$-i\hbar\frac{d}{dx}e^{\pm ikx} = \pm\hbar k e^{\pm ikx}. \tag{2.122}$$

that $-i\hbar\frac{d}{dx} = \pm\hbar k$ is the momentum. The two parts of the wavefunction have the following interpretation:

$$Ae^{ikx} : \quad \text{particle moving to the right with momentum}$$
$$\sqrt{2m(E - V)}$$
$$Be^{-ikx} : \quad \text{particle moving to the left with momentum}$$
$$-\sqrt{2m(E - V)} \tag{2.123}$$

The probability current has the form:

$$j(x,t) = \frac{\hbar}{m}\mathrm{Re}\,\bar{\psi}\,\frac{1}{i}\frac{\partial\psi}{\partial x}$$

$$= \frac{\hbar}{m}\mathrm{Re}\left(k|A|^2 - k|B|^2 + kA\bar{B}e^{2ikx} - k\bar{A}Be^{-2ikx}\right)$$

$$= \frac{\hbar k}{m}\left(|A|^2 - |B|^2\right) \tag{2.124}$$

Note that the plane wave $Ae^{ikx}$ extends over all of $\mathbb{R}$. They cannot be normalized because $|\psi(x)|^2 = |A|^2 = \text{constant}$. Moreover, recall that the notion of probability current was developed in the context of normalized wavefunctions. Nevertheless, we can think of this notion on intervals of finite length. In particular, for our applications we will consider the probability current in a finite interval containing the step, i.e., the point where the potential changes. This enables the probability current to play a role in determining matching conditions for the wavefunction on either side of an obstacle. We will see this in Section 2.8. We also remark that in these applications the wavefunction can be thought of as representing a beam of particles of finite flux $j$.

$\boxed{E < V.}$

In this case we set

$$\tilde{k} = \frac{1}{\hbar}\sqrt{2m(V - E)}, \tag{2.125}$$

and the time independent Schrödinger equation takes the form:

$$\frac{d^2\psi}{dx^2} = \tilde{k}^2\psi, \tag{2.126}$$

which has the solution

$$\psi(x) = Ae^{-\tilde{k}x} + Be^{\tilde{k}x}. \tag{2.127}$$

Hence, the wavefunction has exponentially growing solutions as $|x| \to \infty$ for nonzero $A$ and $B$. This will require careful consideration when choosing an allowable form of the wavefunction in specific settings, as we will see.

### *Step Potentials*

We next consider a potential of the form

$$V(x) = \begin{cases} 0 & x < 0 \\ V & x > 0 \end{cases} \tag{2.128}$$

with $V > 0$.

Solutions of the time independent Schrödinger equation were given in Section 2.8. Matching conditions for the wavefunction at the step are obtained from the continuity equation for the probability current (recall Theorem 6 and Section 2.1):

$$\frac{\partial}{\partial t}|\psi|^2 + \frac{\partial j}{\partial x} = 0. \tag{2.129}$$

Requiring the probability current to be independent of time gives:

$$\frac{\partial j}{\partial x} = 0 \Rightarrow j = \text{constant}. \tag{2.130}$$

Conservation of probability across the step implies that we have the following "matching condition" for the probability current at the step:

$$\frac{\hbar}{m}\text{Re}\left(\bar{\psi}\frac{1}{i}\frac{\partial \psi}{\partial x}\right)\Big|_{x=0-} = \frac{\hbar}{m}\text{Re}\left(\bar{\psi}\frac{1}{i}\frac{\partial \psi}{\partial x}\right)\Big|_{x=0+} \tag{2.131}$$

The matching condition will be satisfied if we require $\psi(x)$ and $\frac{\partial \psi}{\partial x}(x)$ to be continuous across a finite step.

$$\boxed{E > V.}$$

For definiteness, consider particles incident from left to right. Classically all particles continue from $x < 0$ to $x > 0$ with momenta:

$$p = \begin{cases} \sqrt{2mE} & x < 0, \\ \sqrt{2m(E-V)} & x > 0. \end{cases} \tag{2.132}$$

Quantum mechanically, the solution of the time independent Schrödinger equation is given by:

$$\psi(x) = \begin{cases} A_0 e^{ik_0 x} + B_0 e^{-ik_0 x}, & x < 0, \\ A_1 e^{ik_1 x} + B_1 e^{-ik_1 x}, & x > 0, \end{cases} \tag{2.133}$$

where

$$k_0 = \frac{1}{\hbar}\sqrt{2mE} \qquad k_1 = \frac{1}{\hbar}\sqrt{2m(E-V)} \tag{2.134}$$

Since the particles are incoming from left to right there is no component of the wavefunction moving from right to left in $x > 0$. This implies that $B_1 = 0$

Continuity of $\psi(x)$ at $x = 0$ implies

$$A_0 + B_0 = A_1. \tag{2.135}$$

Continuity of $\frac{\partial \psi}{\partial x}(x)$ at $x = 0$ implies

$$k_0(A_0 - B_0) = k_1 A_1. \tag{2.136}$$

Using (2.135) and (2.136), after some algebra we obtain:

$$\begin{aligned}
A_1 &= \frac{2k_0}{k_0+k_1} A_0, \\
B_0 &= A_1 - A_0 = \frac{k_0-k_1}{k_0+k_1} A_0.
\end{aligned} \tag{2.137}$$

We saw in Section 2.8 that for a wavefunction of the form $\psi(x) = Ae^{ikx} + Be^{-ikx}$ the probability current has the form:

$$j = \frac{\hbar k}{m}\left(|A|^2 - |B|^2\right). \tag{2.138}$$

Using the fact that the probability current is continuous across the step, and therefore it has the same value on each side of the step, we obtain:

$$k_0|A_0|^2 - k_0|B_0|^2 = k_1|A_1|^2. \tag{2.139}$$

The terms in (2.139) have the following interpretation:

$$\begin{aligned}
k_0|A_0|^2 &: \quad \text{probability current from left to right for } x < 0, \\
k_0|B_0|^2 &: \quad \text{probability current from right to left for } x < 0, \\
k_1|A_1|^2 &: \quad \text{probability current from left to right for } x > 0.
\end{aligned} \tag{2.140}$$

We rewrite (2.139) as:

$$1 - \frac{|B_0|^2}{|A_0|^2} = \frac{k_1}{k_0}\frac{|A_1|^2}{|A_0|^2}. \tag{2.141}$$

We have the following definitions.

**Definition 23 (Reflection Coefficient).** The reflection coefficient is defined as:

$$R = \frac{|B_0|^2}{|A_0|^2} = \frac{\text{reflected probability current}}{\text{incoming probability current}}. \tag{2.142}$$

**Definition 24 (Transmission Coefficient).**

The transmission coefficient is defined as

$$T = \frac{k_1|A_1|^2}{k_0|A_0|^2} = \frac{\text{transmitted probability current}}{\text{incoming probability current}}. \tag{2.143}$$

Using these definitions, along with the conservation of probability current given in (2.141), we can express the conservation of probability current as:

$$R + T = 1. \tag{2.144}$$

$$\boxed{0 < E < V.}$$

For this case, for a wave incident from left to right, classically all particles are reflected from the step. You can only find particles for $x < 0$. The situation for quantum mechanics is different, as we will see.

The solution of the time independent Schrödinger equation is given by:

$$\psi(x) = \begin{cases} A_0 e^{ik_0 x} + B_0 e^{-ik_0 x}, & x < 0 \\ A_1 e^{-\tilde{k}_1 x} + B_1 e^{\tilde{k}_1 x}, & x > 0 \end{cases} \tag{2.145}$$

where

$$k_0 = \frac{1}{\hbar}\sqrt{2mE}, \qquad \tilde{k}_1 = \frac{1}{\hbar}\sqrt{2m(V - E)}. \tag{2.146}$$

For $\psi(x)$ to remain bounded as $x \to \infty$ we require $B_1 = 0$. The matching conditions at the step for $\psi$ are:

$$A_0 + B_0 = A_1,$$

$$ik_0(A_0 - B_0) = -\tilde{k}_1 A_1. \tag{2.147}$$

Using simple algebraic manipulations, (2.147) can be rewritten as:

$$A_1 = \frac{2k_0}{k_0 + i\tilde{k}_1} A_0 \qquad B_0 = \frac{k_0 - i\tilde{k}_1}{k_0 + i\tilde{k}_1} A_0, \tag{2.148}$$

where the latter expression can be simplified to

$$|B_0| = |A_0|, \tag{2.149}$$

which can be written as:

$$\frac{|B_0|^2}{|A_0|^2} = 1. \tag{2.150}$$

From this relation, and the definition of the reflection coefficient, we have:

$$R = \frac{|B_0|^2}{|A_0|^2} = 1, \tag{2.151}$$

and therefore

$$T = 1 - R = 0 \tag{2.152}$$

Nevertheless, for $x > 0$

$$|\psi|^2 = |A_1|^2 e^{-2\tilde{k}_1 x}. \tag{2.153}$$

Hence there is a finite probability that the particle can be found in $x > 0$. This is related to the quantum mechanical phenomenon of tunneling.

While our analysis shows that probability density can be found in the classically forbidden region, i.e. "inside the barrier", since the barrier is infinite in length, and the probability density decays at an exponential rate with respect to $x$, it can never "get out the other side". Next we will consider the case of a finite barrier where the possibility of probability current being transmitted through the barrier exists. This is referred to as *quantum tunnelling*.

### A Barrier of Finite Width: Quantum Tunnelling

We next consider the case of a barrier of finite width on the line. This is described by the potential

$$V(x) = \begin{cases} 0 & x < 0, \\ V & 0 < x < a, \\ 0 & x > a. \end{cases} \tag{2.154}$$

The particle is incident from left to right in $x < 0$, and we will only consider the case $0 < E < V$. In this case the solution to the time independent Schrödinger equation is given by:

$$\psi(x) = \begin{cases} A_0 e^{ik_0 x} + B_0 e^{-ik_0 x}, & x < 0, \\ A_1 e^{-\tilde{k}_1 x} + B_1 e^{\tilde{k}_1 x}, & 0 < x < a, \\ A_2 e^{ik_0 x} + B_2 e^{-ik_0 x}, & x > a, \end{cases} \tag{2.155}$$

where

$$k_0 = \frac{1}{\hbar}\sqrt{2mE}, \qquad \tilde{k}_1 = \frac{1}{\hbar}\sqrt{2m(V - E)}. \tag{2.156}$$

We note that $B_2 = 0$ since there is no wave moving to the left for $x > a$.

We will compute the reflection coefficient, $R$ from the boundary $x = 0$, and the transmission coefficient, $T$, across the boundary $x = a$. These are given by:

$$R = \frac{|B_0|^2}{|A_0|^2}, \quad T = \frac{|A_2|^2}{|A_0|^2}. \tag{2.157}$$

We apply the matching conditions at the boundaries to (2.155) in order to obtain relations amongst the coefficients. The matching conditions give:

$$\text{at } x = 0, \quad \begin{aligned} \psi : \quad & A_0 + B_0 = A_1 + B_1, \quad \text{(i)}, \\ \frac{d\psi}{dx} : \quad & ik_0(A_0 - B_0) = -\tilde{k}(A_1 - B_1), \quad \text{(ii)}, \end{aligned} \tag{2.158}$$

$$\text{at } x = a, \quad \begin{aligned} \psi : \quad & A_1 e^{-\tilde{k}_1 a} + B_1 e^{\tilde{k}_1 a} = A_2 e^{ik_0 a}, \quad \text{(iii)}, \\ \frac{d\psi}{dx} : \quad & -\tilde{k}_1 \left( A_1 e^{-\tilde{k}_1 a} - B_1 e^{\tilde{k}_1 a} \right) = ik_0 A_2 e^{ik_0 a}, \quad \text{(iv)}. \end{aligned} \tag{2.159}$$

It is clear from the form of (2.157) that if we can express $B_0 = \text{constant} A_0$ and $A_2 = \text{constant}' A_0$, where constant and constant$'$ are functions of the system parameters $a$, $k_0$, $\tilde{k}_1$ then expressions for the reflection and transmission coefficients in terms of these same system parameters easily follow. After some algebra, we obtain:

$$A_1 = \frac{1}{2}\left( 1 - \frac{ik_0}{\tilde{k}_1} \right) A_2 e^{ik_0 a} e^{\tilde{k}_1 a} \qquad B_1 = \frac{1}{2}\left( 1 + \frac{ik_0}{\tilde{k}_1} \right) A_2 e^{ik_0 a} e^{-\tilde{k}_1 a} \tag{2.160}$$

$$\begin{aligned} A_0 &= \frac{1}{2} A_1 \left( 1 - \frac{\tilde{k}_1}{ik_0} \right) + \frac{1}{2} B_1 \left( 1 + \frac{\tilde{k}_1}{ik_0} \right), \\ &= \frac{1}{4}\left( 1 - \frac{ik_0}{\tilde{k}_1} + 1 + \frac{i\tilde{k}_1}{k_0} \right) A_2 e^{ik_0 a} e^{\tilde{k}_1 a} + \frac{1}{4}\left( 1 + \frac{ik_0}{\tilde{k}_1} + 1 - \frac{i\tilde{k}_1}{k_0} \right) \\ &\quad \times A_2 e^{ik_0 a} e^{-\tilde{k}_1 a}, \\ &= \left( \cosh(\tilde{k}_1 a) + \frac{i}{2}\left( \frac{\tilde{k}_1}{k_0} - \frac{k_0}{\tilde{k}_1} \right) \sinh(\tilde{k}_1 a) \right) A_2 e^{ik_0 a}, \end{aligned} \tag{2.161}$$

$$\begin{aligned} B_0 &= \frac{1}{2} A_1 \left( 1 + \frac{\tilde{k}_1}{ik_0} \right) + \frac{1}{2} B_1 \left( 1 - \frac{\tilde{k}_1}{ik_0} \right) = \frac{i}{2}\left( -\frac{\tilde{k}_1}{k_0} - \frac{k_0}{\tilde{k}_1} \right) \\ &\quad \times \sinh(\tilde{k}_1 a) A_2 e^{ik_0 a}. \end{aligned} \tag{2.162}$$

Therefore the reflection coefficient is given by

$$R = \frac{|B_0|^2}{|A_0|^2} = \frac{\dfrac{(\tilde{k}_1^2 + k_0^2)^2}{4k_0^2\tilde{k}_1^2}\sinh^2(\tilde{k}_1 a)}{\cosh^2(\tilde{k}_1 a) + \dfrac{(\tilde{k}_1^2 - k_0^2)^2)}{4k_0^2\tilde{k}_1^2}\sinh^2(\tilde{k}_1 a)}$$

$$= \frac{\dfrac{(\tilde{k}_1^2 + k_0^2)^2}{4k_0^2\tilde{k}_1^2}\sinh^2(\tilde{k}_1 a)}{1 + \dfrac{(\tilde{k}_1^2 + k_0^2)^2}{4k_0^2\tilde{k}_1^2}\sinh^2(\tilde{k}_1 a)}, \tag{2.163}$$

and the transmission coefficient is given by

$$T = \frac{|A_2|^2}{|A_0|^2} = \frac{1}{1 + \frac{(\tilde{k}_1^2 + k_0^2)^2}{4k_0^2\tilde{k}_1^2}\sinh^2(\tilde{k}_1 a)}. \tag{2.164}$$

Hence we have quantum transmission, even though classically it is impossible. This is called *quantum tunneling*. As a check on these calculations we easily see that $R + T = 1$.

It is insightful to consider some limiting cases. For $\tilde{k}_1 a$ "very large" we have

$$T \approx \frac{4k_0^2\tilde{k}_1^2}{(\tilde{k}_1^2 + k_0^2)^2}e^{-2\tilde{k}_1 a}. \tag{2.165}$$

Hence the transmission coefficient becomes "small", and therefore reflection dominates. Alternately, for $\tilde{k}_1 a$ "very small" the transmission coefficient takes the form

$$T \approx 1 - \frac{(\tilde{k}_1^2 + k_0^2)^2}{4k_0^2}a^2, \tag{2.166}$$

and transmission dominates over reflection.

### Problems

1. This exercise is from the book by Hannabuss.[22] A particle of mass $m$ moves freely in the interval $[0, a]$ on the $x$-axis. The wavefunction describing it may be assumed to vanish at $x = 0$ and $x = a$. Initially the wavefunction is given by:

$$\psi(x, 0) = \frac{1}{\sqrt{a}}\sin\left(\frac{\pi x}{a}\right)\left(1 + 2\cos\left(\frac{\pi x}{a}\right)\right). \tag{2.167}$$

---

[22]K. Hannabuss. *An introduction to quantum theory*, volume 1. Clarendon Press, 1997.

Show that at a later time $t$, the wavefunction is given by:

$$\psi(x,t) = \frac{1}{\sqrt{a}} \exp\left(-\frac{i\pi^2 \hbar t}{2ma^2}\right) \sin\left(\frac{\pi x}{a}\right)$$

$$\times \left(1 + 2\exp\left(-\frac{i3\pi^2 \hbar t}{2ma^2}\right) \cos\left(\frac{\pi x}{a}\right)\right). \tag{2.168}$$

Find the probability that at time $t$ the particle lies in the interval $[0, a/2]$. What is the probability that the particle's energy will be measured to be that of the $n^{\text{th}}$ energy level of the system?

2. Repeat the calculations in Problem 1 for the case when the initial condition is given by:

$$\psi(x,0) = \sqrt{\frac{12}{a^3}} \begin{cases} x & \text{if } \ 0 < x \leq \dfrac{a}{2} \\ a - x & \text{if } \ \dfrac{a}{2} \leq x < a, \end{cases} \tag{2.169}$$

and verify that the wavefunction is normalized.

3. This exercise is from the book by Hannabuss.[23] A particle of mass $m$, moving freely between impenetrable barriers at $x = 0$ and $x = a$ (you may assume that quantum wavefunctions vanish at such barriers), is in the eigenstate corresponding to the lowest energy level when the barrier at $x = a$ is suddenly displaced to $x = 2a$, at time $t = 0$. By expanding the original wavefunction in terms of the eigenfunctions for motion within $[0, 2a]$, find the wavefunction at a subsequent time $t$, and show that it is a superposition of states of energies

$$E = \frac{n^2 \pi^2 \hbar^2}{8ma^2}. \tag{2.170}$$

for $n = 2$ and $n = 1, 3, 5, \ldots$ Show that the probability of finding the particle's energy to be unchanged is $\frac{1}{2}$.

4. Consider the time-independent Schrödinger equation

$$-\frac{\hbar^2}{2m} \frac{d^2\psi}{dx^2}(x) + V(x)\psi(x) = E\psi(x), \tag{2.171}$$

with the boundary condition that $\psi(x) \to 0$ as $|x| \to \infty$, sufficiently fast for the solutions to be normalizable. Prove that the energy levels are all non-degenerate.

---

[23] K. Hannabuss. *An introduction to quantum theory*, volume 1. Clarendon Press, 1997.

5. Let

$$H = -\frac{\hbar^2}{2m}\frac{d^2}{dx^2} + V(x),\qquad(2.172)$$

and $\alpha(x)$ and $\beta(x)$ be any two functions that are twice differentiable and satisfy $\alpha(x) \to 0$ and $\beta(x) \to 0$ as $|x| \to \infty$. Prove that

$$\int_{-\infty}^{\infty} \overline{\alpha} H\beta \, dx = \int_{-\infty}^{\infty} \beta H\overline{\alpha} \, dx.\qquad(2.173)$$

6. Let $\psi(x,t)$ be a solution of the Schrödinger equation

$$H\psi = i\hbar\frac{\partial\psi}{\partial t},\qquad(2.174)$$

where $H$ is as in (2.172) and $V(x) \to 0$ as $|x| \to \infty$. Show that the expectation value of the momentum for a particle in the state represented by $\psi$ is given by

$$\langle p \rangle = -i\hbar \int_{-\infty}^{\infty} \overline{\psi}\frac{\partial\psi}{\partial x} \, dx.\qquad(2.175)$$

7. Let $\psi(x,t)$ be a solution of the Schrödinger equation in the previous problem. Show that for operators $A$ that do not depend explicitly on time $t$,

$$i\hbar\frac{d}{dt}\int_{-\infty}^{\infty} \overline{\psi}(x,t)A\psi(x,t) \, dx = \int_{-\infty}^{\infty} \overline{\psi}(x,t)[A,H]\psi(x,t) \, dx,\quad(2.176)$$

where $[A,H] = AH - HA$. Hence prove that

(a) the expectation value of the energy is a constant of the (quantum) dynamics when the potential $V$ is time-independent;

(b)
$$\frac{d}{dt}\langle p \rangle = -\int_{-\infty}^{\infty} \overline{\psi}(x,t)\frac{dV}{dx}\psi(x,t) \, dx,$$

$$= -\left\langle \frac{dV}{dx} \right\rangle.\qquad(2.177)$$

This result is known as *Ehrenfest's Theorem*. Comment on the relationship of 1 and 2 with the corresponding results in classical mechanics.

8. A particle of mass $m$ moves freely in the interval $[0, a]$ on the $x$-axis. The wavefunction describing it may be assumed to vanish at $x = 0$ and $x = a$. Initially the wavefunction is given by:

$$\psi(x, 0) = \frac{1}{\sqrt{a}} \sin\left(\frac{\pi x}{a}\right)\left(1 + 2\cos\left(\frac{\pi x}{a}\right)\right). \tag{2.178}$$

In problem 1 we showed that at a later time $t$ the wavefunction is given by:

$$\psi(x, t) = \frac{1}{\sqrt{a}} \exp\left(-\frac{i\pi^2 \hbar t}{2ma^2}\right) \sin\left(\frac{\pi x}{a}\right)$$

$$\times \left(1 + 2\exp\left(-\frac{i3\pi^2 \hbar t}{2ma^2}\right)\cos\left(\frac{\pi x}{a}\right)\right). \tag{2.179}$$

Compute the expectation value of the energy in this state, first by calculating

$$i\hbar \int_0^a \overline{\psi}\frac{\partial \psi}{\partial t}\,dx, \tag{2.180}$$

and second by using the probability for the particle to be found in the $n^{\text{th}}$ energy eigenstate of the system.

9. Particles with energy $E > 0$, quantum mechanically described by a wave incident from $-\infty$, move along the $x$-axis in the potential

$$V(x) = \begin{cases} 0 & \text{if } x < 0, \\ -V & \text{if } x > 0 \end{cases} \tag{2.181}$$

where $V > 0$. Find the reflection and transmission coefficients, $R$ and $T$. Show that, contrary to the classical result, $R \to 1$ in the limit as $\frac{V}{E} \to \infty$.

10. Particles with energy $E$, quantum mechanically described by a wave incident from $-\infty$, move along the $x$-axis in the potential

$$V(x) = \begin{cases} 0 & \text{if } x < 0, \\ V & \text{if } 0 < x < a, \\ 0 & x > a \end{cases} \tag{2.182}$$

where $E > V > 0$. Find the reflection and transmission coefficients, $R$ and $T$. Show that when

$$\sqrt{2m(E - V)} = n\pi\hbar, \tag{2.183}$$

where $n = 0, 1, 2, \ldots$ then $T = 1$ (and hence $R = 0$).

11. Consider a particle in the infinite square well of width 1 (the potential is $V(x) = 0$ on $[0,1]$ and $\infty$ everywhere else). Recall that the time-independent Schrödinger equation has eigenvalues at

$$E_n = \frac{n^2\pi^2\hbar^2}{2m},$$

with associated wavefunctions

$$\psi_n(x) = \sqrt{2}\sin(n\pi x).$$

Suppose that the particle in this well has as its initial wave function an even mixture of the first two stationary states:

$$\Psi(x,0) = A(\psi_1(x) + \psi_2(x)),$$

for some normalization constant $A$.

(a) Normalize $\Psi(x,0)$.

(b) Find $\Psi(x,t)$ and $|\Psi(x,t)|^2$. Express the latter as a sinusoidal function of time. To simplify the result, let $\omega = \pi^2\hbar/2m$.

(c) Compute $\langle x \rangle$.

   *Hint.* Recall that

$$\cos^2(\theta/2) = \frac{1 + \cos\theta}{2},$$

   and integrate by parts. Furthermore,

$$\sin(\pi x)\sin(2\pi x) = \frac{1}{2}(\cos(\pi x) - \cos(3\pi x)).$$

(d) Compute $\langle p \rangle$.

(e) If you measured the energy of this particle, what values might you get, and what is the probability of getting each of them? Find the expectation value of $H$.

12. Consider a quantum particle of constant mass $m$ moving in a potential well $V(x)$, $x \in \mathbb{R}$, whose initial state, $\Psi(x,0)$, is given by an equal weight superposition of the ground state, $\psi_1(x)$, and the first excited state, $\psi_2(x)$, i.e.,

$$\Psi(x,0) = C(\psi_1(x) + \psi_2(x)),$$

where $\psi_1(x)$ and $\psi_2(x)$ are normalized and real.

(a) Determine $C$ so that $\Psi(x,0)$ is normalized.

(b) Determine $\Psi(x,t)$ for any later time $t$.

(c) Show that the average energy, $\langle E \rangle$, for $\Psi(x, t)$ is the arithmetic mean of the energies of the ground state and the first excited state energies $E_1$ and $E_2$, respectively. That is $\langle E \rangle = \frac{1}{2}\left(E_1 + E_2\right)$.

(d) Determine the uncertainty of energy, $\Delta E$, for $\Psi(x, t)$.

(e) Derive a time periodic function for the average position $\langle x \rangle$ in the state $\Psi(x, t)$.

13. Consider the one dimensional time independent Schrödinger equation:

$$-\frac{\hbar^2}{2m}\frac{d^2\psi}{dx^2}(x) + V(x)\psi(x) = E\psi(x), \qquad (2.184)$$

with the boundary condition $\psi(x) \to 0$ as $|x| \to \infty$, sufficiently fast for the solutions to be normalizable. We assume that $V(x)$ is bounded on $\mathbb{R}$ and we denote the energy eigenstates and eigenvalues by $(\psi_n(x), E_n)$.

(a) Prove that the energy levels are all non-degenerate. (Hint: argue by contradiction.)

(b) Show that the energy eigenstates can be chosen to be real. (Hint: use part (a) of this problem.)

(c) Let $\psi$ be a solution of the one dimensional time independent Schrödinger equation. Show that

$$\frac{\hbar}{2mi}\left(\overline{\psi}\psi' - \psi\overline{\psi}'\right) = \frac{1}{m}\mathrm{Re}\left(\overline{\psi}\frac{\hbar}{i}\psi'\right),$$

where $\psi'$ denotes the derivative of $\psi$.

# Chapter 3

# Measurement, Time Evolution, Uncertainty, and the Harmonic Oscillator

We will use the mathematical structure developed in the first chapter to study the quantum theory of measurement, uncertainty, and time evolution. We will conclude this chapter with a study of one of the fundamental quantum mechanical systems–the harmonic oscillator.

## 3.1  Measurement

Now we will describe some aspects of the "quantum theory of measurement". This will play an important role in the remainder of the book. We begin by recalling two facts that were mentioned earlier.

- The state of a physical system is described by a normalised ket, $|\psi\rangle$, in a Hilbert space $\mathcal{H}$.
- Every measurable physical quantity of a physical system ("observable") is described by a self-adjoint operator, $A$, acting on the Hilbert space $\mathcal{H}$.

We now give a brief description of the three "rules" or postulates of measurement. Often these, collectively go by the name of the "Born rule", since they were formulated in a paper of Max Born published in 1926.

*Zur Quantenmechanik der Stoßvorgänge, Max Born, Zeitschrift für Physik, 37, #12 (Dec. 1926), pp. 863–867 (German); English translation, On the quantum mechanics of collisions, in Quantum theory and measurement, section I.2, J. A. Wheeler and W. H. Zurek, eds., Princeton, NJ: Princeton University Press, 1983, ISBN 0-691-08316-9.*

**Outcome of a Measurement.** The only possible outcome of the measurement of a physical observable $A$ is an eigenvalue of $A$, $\lambda$.

**Probability for a Particular Outcome of a Measurement.**
Suppose the system is in the state, $|\psi\rangle$, and the observable $A$ is measured. Then the eigenvalue of $A$, $\lambda$, occurs with the probability:

$$\mathrm{prob}_\psi(\lambda) \equiv \langle\psi|P_\lambda|\psi\rangle = \||P_\lambda|\psi\rangle\|^2, \tag{3.1}$$

where $P_\lambda$ is the orthogonal projection onto the subspace spanned by the eigenvectors corresponding to $\lambda$.

**State of the System After Measurement.** If the measurement of $A$ on the system that is in the state $|\psi\rangle$ gives the outcome $\lambda$, then the state of the system after the measurement is given by:

$$\frac{P_\lambda|\psi\rangle}{\sqrt{\langle\psi|P_\lambda|\psi\rangle}}, \tag{3.2}$$

where, recall

$$\||P_\lambda|\psi\rangle\| = \sqrt{\langle\psi|P_\lambda^\dagger P_\lambda|\psi\rangle} = \sqrt{\langle\psi|P_\lambda^2|\psi\rangle} = \sqrt{\langle\psi|P_\lambda|\psi\rangle}. \tag{3.3}$$

**Summary:** *Measurement of an observable yields an eigenvalue. If we know the state before the measurement, we can calculate the probability of obtaining a particular eigenvalue as a result of measurement in that state. However, the measurement changes the state, but we can calculate that state after the measurement, provided we know the state in which the measurement is made, and the outcome of the measurement.*

*Examples.* We now consider some examples that illustrate the calculations and formalism in the quantum theory of measurement.

Let $\lambda_i$ and $|e_i\rangle$ denoted the eigenvalues and (normalised) eigenvectors of a self-adjoint linear operator $A$ defined on a finite dimensional Hilbert space. Let us suppose that there are no degeneracies, then the eigenspace corresponding to an eigenvalue $\lambda_i$ is one dimensional and the associated projection operators are given by:

$$P_{\lambda_i} = |e_i\rangle\langle e_i|, \tag{3.4}$$

and $A$ can be represented in the following spectral form:

$$A = \sum_i \lambda_i |e_i\rangle\langle e_i|, \quad \langle e_i|e_i\rangle = 1. \tag{3.5}$$

- Suppose the system is in the *normalised* state:

$$|\psi\rangle = \sum_i a_i |e_i\rangle, \tag{3.6}$$

and suppose $A$ is measured in the state $|\psi\rangle$. Then, using (3.1), the probability that the outcome of the measurement of $A$ in $|\psi\rangle$ is $\lambda_i$ is given by:

$$\text{prob}_\psi(\lambda_i) = \langle\psi|P_{\lambda_i}|\psi\rangle = \langle\psi|e_i\rangle\langle e_i|\psi\rangle = |\langle e_i|\psi\rangle|^2 = |a_i|^2. \tag{3.7}$$

Using (3.2), if $\lambda_i$ is measured, then after the measurement the state is given by:

$$\frac{P_{\lambda_i}|\psi\rangle}{\sqrt{\langle\psi|P_{\lambda_i}|\psi\rangle}} = \frac{|e_i\rangle\langle e_i|\psi\rangle}{|\langle e_i|\psi\rangle|} = \frac{\langle e_i|\psi\rangle}{|\langle e_i|\psi\rangle|}|e_i\rangle = \frac{a_i}{|a_i|}|e_i\rangle. \tag{3.8}$$

- Note that we have:

$$\sum_i \text{prob}_\psi(\lambda_i) = \sum_{\lambda_i}\langle\psi|P_{\lambda_i}|\psi\rangle = \sum_i \langle\psi|e_i\rangle\langle e_i|\psi\rangle,$$

$$= \langle\psi|\left(\sum_i |e_i\rangle\langle e_i|\right)|\psi\rangle = \langle\psi|\psi\rangle = 1. \tag{3.9}$$

- Suppose $|\psi\rangle$ is an eigenstate of $A$, i.e.,

$$|\psi\rangle = |e_n\rangle.$$

Then we have

$$\text{prob}_\psi(\lambda_n) = |\langle e_n|\psi\rangle|^2 = 1,$$

and

$$\text{prob}_\psi(\lambda_m) = |\langle e_m|\psi\rangle|^2 = 0, \quad m \neq n.$$

In other words, if we measure $A$ in an eigenstate of $A$ the outcome of the measurement is always the eigenvalue corresponding to the eigenstate.

- Recall that the *expectation value of $A$ in the state* $|\psi\rangle$, denoted $E_\psi(A) \equiv \langle\psi|A|\psi\rangle$, is the predicted *mean value* of the measurement of $A$ in the state $\psi$. Using (3.5), we have:

$$E_\psi(A) = \langle\psi|A|\psi\rangle,$$

$$= \sum_i \lambda_i \langle\psi|e_i\rangle\langle e_i|\psi\rangle,$$

$$= \sum_i \lambda_i \langle\psi|P_{\lambda_i}|\psi\rangle,$$

$$= \sum_i \lambda_i \text{prob}_\psi(\lambda_i). \tag{3.10}$$

- The *dispersion of an operator $A$ in the state $|\psi\rangle$* is defined as follows:

$$\Delta_\psi(A) = \left[ E_\psi\left( (A - E_\psi(A))^2 \right) \right]^{\frac{1}{2}},$$

$$= \left[ \langle\psi|(A^2 - 2A\langle\psi|A|\psi\rangle + \langle\psi|A|\psi\rangle^2)|\psi\rangle \right]^{\frac{1}{2}},$$

$$= \left[ \langle\psi|A^2|\psi\rangle - \langle\psi|A|\psi\rangle^2 \right]^{\frac{1}{2}},$$

$$= \left[ E_\psi(A^2) - (E_\psi(A))^2 \right]^{\frac{1}{2}}. \tag{3.11}$$

This quantity will play an important role when we discuss the *generalised uncertainty relation* a bit later.

*It is important to keep in mind that the expectation value of an operator and the dispersion of an operator, in general, depend on the state $|\psi\rangle$.*

*Example.* Suppose $\mathcal{H} = \mathbb{C}^2$, and let

$$\{|1\rangle,\, |2\rangle\} = \left\{ \begin{pmatrix} 1 \\ 0 \end{pmatrix},\ \begin{pmatrix} 0 \\ 1 \end{pmatrix} \right\}$$

denote an orthonormal basis of $\mathbb{C}^2$. We define a self adjoint linear operator on $\mathbb{C}^2$ as follows:

$$A|1\rangle = 4|1\rangle - 2|2\rangle$$

$$A|2\rangle = -2|1\rangle + 4|2\rangle. \tag{3.12}$$

In the basis $\{|1\rangle,\, |2\rangle\}$ the matrix representation of this linear operator is given by:

$$A = \begin{pmatrix} 4 & -2 \\ -2 & 4 \end{pmatrix}, \tag{3.13}$$

and we easily verify that $A = A^\dagger$.

We can also express this linear map in Dirac notation in the basis $\{|1\rangle,\, |2\rangle\}$ as follows:

$$A = 4|1\rangle\langle1| - 2|1\rangle\langle2| - 2|2\rangle\langle1| + 4|2\rangle\langle2|. \tag{3.14}$$

The eigenvalues of $A$ are determined from:

$$\det \begin{pmatrix} 4 - \lambda & -2 \\ -2 & 4 - \lambda \end{pmatrix} \tag{3.15}$$

or

$$(4 - \lambda)^2 - 4 = 0. \tag{3.16}$$

From which it follows that:

$$\lambda_1 = 6, \quad \lambda_2 = 2,$$

with the (normalised) eigenvectors corresponding to these eigenvalues given by:

$$|e_1\rangle = \frac{1}{\sqrt{2}} \begin{pmatrix} 1 \\ -1 \end{pmatrix}, \quad |e_2\rangle = \frac{1}{\sqrt{2}} \begin{pmatrix} 1 \\ 1 \end{pmatrix}$$

By inspection, one can see that $|e_1\rangle$ and $|e_2\rangle$ can be expressed in terms of the basis vectors $\{|1\rangle, |2\rangle\}$ as follows:

$$|e_1\rangle = \frac{|1\rangle - |2\rangle}{\sqrt{2}}, \quad |e_2\rangle = \frac{|1\rangle + |2\rangle}{\sqrt{2}}.$$

Then it is straightforward to represent $A$ in Dirac notation in either basis as follows:

$$A = 6|e_1\rangle\langle e_1| + 2|e_2\rangle\langle e_2|,$$

$$= 6 \left( \frac{|1\rangle - |2\rangle}{\sqrt{2}} \right) \left( \frac{\langle 1| - \langle 2|}{\sqrt{2}} \right) + 2 \left( \frac{|1\rangle + |2\rangle}{\sqrt{2}} \right) \left( \frac{\langle 1| + \langle 2|}{\sqrt{2}} \right),$$

$$= 4|1\rangle\langle 1| - 2|1\rangle\langle 2| - 2|2\rangle\langle 1| + 4|2\rangle\langle 2|. \tag{3.17}$$

Suppose the system is in the state $|1\rangle$, and we wish to measure $A$ in this state. We know that the only possibility for the outcome of a measurement will be an eigenvalue of $A$. Using (3.1), the probabilities for the two possible outcomes are given by:

$$\text{prob}_{|1\rangle} (\lambda = 6) = \langle 1|P_{\lambda_1}|1\rangle,$$

$$= \langle 1|e_1\rangle\langle e_1|1\rangle,$$

$$= \langle 1| \left( \frac{|1\rangle - |2\rangle}{\sqrt{2}} \right) \left( \frac{\langle 1| - \langle 2|}{\sqrt{2}} \right) |1\rangle,$$

$$= \frac{1}{2}, \tag{3.18}$$

and

$$\text{prob}_{|1\rangle}\,(\lambda = 2) = \langle 1|P_{\lambda_2}|1\rangle,$$
$$= \langle 1|e_2\rangle\langle e_2|1\rangle,$$
$$= \langle 1|\left(\frac{|1\rangle + |2\rangle}{\sqrt{2}}\right)\left(\frac{\langle 1| + \langle 2|}{\sqrt{2}}\right)|1\rangle,$$
$$= \frac{1}{2}. \tag{3.19}$$

Note that the probabilities add to one (as they should).

Using (3.10), the expectation value of $A$ in the state $|1\rangle$ is given by:

$$E_{|1\rangle}(A) = \langle 1|A|1\rangle$$
$$= \langle 1|\left(4|1\rangle\langle 1| - 2|1\rangle\langle 2| - 2|2\rangle\langle 1| + 4|2\rangle\langle 2|\right)|1\rangle = 4.$$
$$\tag{3.20}$$

Using (3.11), the dispersion of $A$ in the state $|1\rangle$ is given by:

$$\Delta_{|1\rangle}(A) = \left[\langle 1|A^2|1\rangle - \langle 1|A|1\rangle^2\right]^{\frac{1}{2}}. \tag{3.21}$$

Now we compute the various terms that go into this quantity. From (3.13) we have:

$$A^2 = \begin{pmatrix} 20 & -16 \\ -16 & 20 \end{pmatrix}, \tag{3.22}$$

which in Dirac notation is:

$$A^2 = 20|1\rangle\langle 1| - 16|1\rangle\langle 2| - 16|2\rangle\langle 1| + 20|2\rangle\langle 2|, \tag{3.23}$$

and therefore;

$$\langle 1|A^2|1\rangle = \langle 1|\left(20|1\rangle\langle 1| - 16|1\rangle\langle 2| - 16|2\rangle\langle 1| + 20|2\rangle\langle 2|\right)|1\rangle = 20.$$

Assembling the different quantities, we obtain:

$$\Delta_{|1\rangle}(A) = \left[\langle 1|A^2|1\rangle - \langle 1|A|1\rangle^2\right]^{\frac{1}{2}} = (20 - 16)^{\frac{1}{2}} = 2. \tag{3.24}$$

If the result of the measurement of $A$ in the state $|1\rangle$ is $\lambda = 6$ then, using (3.2), the state of the system after the measurement is given by:

$$\frac{|e_1\rangle\langle e_1|1\rangle}{|\langle e_1|1\rangle|} = |e_1\rangle\frac{\frac{1}{\sqrt{2}}}{\frac{1}{\sqrt{2}}} = |e_1\rangle$$

## *Interference*

Interference is a unique characteristic of quantum systems that illustrates the wave-like nature of matter. We will illustrate this characteristic in the context of the example above.

Suppose we measure $A$ in the state $|\psi\rangle$. If $|\psi\rangle = |1\rangle$, then we showed that the probability that we measure $\lambda = 6$ in the state $|1\rangle$ is $\frac{1}{2}$. In summary:

$$|\psi\rangle = |1\rangle \Rightarrow \mathrm{prob}_{|1\rangle}(\lambda = 6) = \frac{1}{2},$$

Now if $|\psi\rangle = |2\rangle$ then the probability that we measure $\lambda = 6$ is given by:

$$\mathrm{prob}_{|2\rangle}(\lambda = 6) = \langle 2|e_1\rangle\langle e_1|2\rangle = \langle 2| \left(\frac{|1\rangle - |2\rangle}{\sqrt{2}}\right) \left(\frac{\langle 1| - \langle 2|}{\sqrt{2}}\right) |2\rangle.$$

$$= \frac{1}{2}.$$

Now suppose that the system is in the state:

$$|\psi\rangle = \frac{|1\rangle + |2\rangle}{\sqrt{2}},$$

i.e., a linear superposition of $|1\rangle$ and $|2\rangle$. Then let us compute the probability to measure $\lambda = 6$ in the state $|\psi\rangle$:

$$\begin{aligned}
\mathrm{prob}_{|\psi\rangle}(\lambda = 6) &= \langle \psi|P_{\lambda=6}|\psi\rangle, \\
&= \langle \psi|e_1\rangle\langle e_1|\psi\rangle, \\
&= \left(\frac{\langle 1| + \langle 2|}{\sqrt{2}}\right) \left(\frac{|1\rangle - |2\rangle}{\sqrt{2}}\right) \left(\frac{\langle 1| - \langle 2|}{\sqrt{2}}\right) \\
&\times \left(\frac{|1\rangle + |2\rangle}{\sqrt{2}}\right) = 0.
\end{aligned} \tag{3.25}$$

Since the new state $|\psi\rangle$ is a (normalised) linear superposition of $|1\rangle$ and $|2\rangle$, with equal amplitudes, we might "guess" that the probability for measuring $\lambda = 6$ in this state is the sum of the probabilities for measuring $\lambda = 6$ in each state of the linear superposition. However, the calculation (3.25) shows that probabilities "do not add" in this way, i.e. there is destructive interference that is indicative of the "wave-like" characteristic of quantum mechanics.

More generally, suppose we consider the state

$$|\psi\rangle = \frac{|1\rangle + e^{i\theta}|2\rangle}{\sqrt{2}},$$

and we compute the probability to measure $\lambda = 6$ in this state. Following the same procedure as above, this is found to be:

$$
\begin{aligned}
\text{prob}_{|\psi\rangle}(\lambda = 6) &= \langle\psi|P_{\lambda=6}|\psi\rangle, \\
&= \langle\psi|e_1\rangle\langle e_1|\psi\rangle, \\
&= \left(\frac{\langle 1| + e^{-i\theta}\langle 2|}{\sqrt{2}}\right)\left(\frac{|1\rangle - |2\rangle}{\sqrt{2}}\right)\left(\frac{\langle 1| - \langle 2|}{\sqrt{2}}\right) \\
&\quad \times \left(\frac{|1\rangle + e^{i\theta}|2\rangle}{\sqrt{2}}\right), \\
&= \frac{\left(1 - e^{-i\theta}\right)\left(1 - e^{i\theta}\right)}{4}, \\
&= \frac{1 - \cos\theta}{2},
\end{aligned}
\tag{3.26}
$$

which indicates that the *relative* phase between $|1\rangle$ and $|2\rangle$ is observable. However, a "global" phase is not observable, i.e. if we considered the state:

$$|\psi\rangle = e^{i\theta}\left(\frac{|1\rangle + |2\rangle}{\sqrt{2}}\right),$$

and computed the probability to measure $\lambda = 6$ in this state, we would get the same answer as (3.25), i.e. zero.

## 3.2   Unitary Operators, Time Evolution

Unitary operators play a fundamental role in quantum mechanics, as we now discuss in this section. We begin with the definition.

**Definition 25 (Unitary Operator).** Let $U$ be a linear operator on a Hilbert space $\mathcal{H}$. The $U$ is said to be a unitary operator if:

$$U^\dagger U = UU^\dagger = \mathbb{I},$$

where $\mathbb{I}$ is the identity operator on $\mathcal{H}$.

An important property of unitary operators is that they "preserve the inner product". This means the following:

$$(U|\psi\rangle, U|\phi\rangle) = \langle\psi|U^\dagger U|\phi\rangle = \langle\psi|\phi\rangle.$$

The significance of unitary operators in quantum mechanics is that they "govern time evolution of the wave function" in the sense of the following calculations. Let

$$U(t) = \exp\left(-\frac{itH}{\hbar}\right), \tag{3.27}$$

where, generally, the exponential of an operator is defined through the usual series expansion:

$$\exp A = \sum_{n=0}^{\infty} \frac{A^n}{n!}. \tag{3.28}$$

(We leave it as an exercise to show that (3.27) is unitary.) Of course, one must prove that the exponential series defining a unitary operator converges. This is "relatively easy" in finite dimensions since all finite dimensional operators are bounded. Of course, things are more technical in the infinite dimensional case, but we will not consider these issues in this book.

Now we want to show that the time evolution of state vectors is governed by (3.27). Consider

$$|\psi(t)\rangle = U(t)|\psi_0\rangle, \quad \text{for some constant} \quad |\psi_0\rangle \in \mathcal{H}. \tag{3.29}$$

Now we show that $|\psi(t)\rangle$ satisfies the Schrödinger equation:

$$i\hbar\frac{d}{dt}|\psi(t)\rangle = H|\psi(t)\rangle. \tag{3.30}$$

First, note that:

$$\begin{aligned}
\frac{d}{dt}U(t) &= \frac{d}{dt}\exp\left(-\frac{iHt}{\hbar}\right), \\
&= \frac{d}{dt}\sum_{n=0}^{\infty}\left(\frac{-it}{\hbar}\right)^n\frac{H^n}{n!}, \\
&= \sum_{n=1}^{\infty}\left(\frac{-i}{\hbar}H\right)\left(\frac{-it}{\hbar}\right)^{n-1}n\frac{H^{n-1}}{n!}, \\
&= \left(\frac{-iH}{\hbar}\right)\sum_{n=1}^{\infty}\left(\frac{-itH}{\hbar}\right)^{n-1}\frac{1}{(n-1)!}, \\
&= \left(-\frac{iH}{\hbar}\right)\exp\left(-\frac{iHt}{\hbar}\right), \\
&= \left(-\frac{iH}{\hbar}\right)U(t). \tag{3.31}
\end{aligned}$$

Then we have:

$$i\hbar\frac{d}{dt}|\psi(t)\rangle = i\hbar\left(\frac{d}{dt}U(t)\right)|\psi_0\rangle + i\hbar U(t)\frac{d}{dt}|\psi_0\rangle,$$

$$= i\hbar\frac{-iH}{\hbar}U(t)|\psi_0\rangle$$

$$= HU(t)|\psi_0\rangle,$$

$$= H|\psi(t)\rangle, \tag{3.32}$$

which shows that an initial condition $|\psi_0\rangle \in \mathcal{H}$ for the Schrödinger equation evolves in time by application of the unitary operator, $|\psi(t)\rangle = U(t)|\psi_0\rangle = \exp\left(-\frac{itH}{\hbar}\right)|\psi_0\rangle$.

Finally, note that if $|e_i\rangle$ is an eigenstate of $H$ with eigenvalue $E_i$, then we have:

$$U(t)|e_i\rangle = \exp\left(-\frac{itH}{\hbar}\right)|e_i\rangle = \exp\left(-\frac{itE_i}{\hbar}\right)|e_i\rangle. \tag{3.33}$$

(This really requires a proof, using the exponential series expression for a unitary operator, that you should be able to provide.) So if we express $|\psi_0\rangle$ in a basis of eigenstates of $H$:

$$|\psi_0\rangle = \sum_i a_i|e_i\rangle, \tag{3.34}$$

Then we have:

$$|\psi(t)\rangle = \exp\left(-\frac{itH}{\hbar}\right)\sum_i a_i|e_i\rangle = \sum_i a_i \exp\left(-\frac{itE_i}{\hbar}\right)|e_i\rangle. \tag{3.35}$$

### Commutation Relations

In general, two linear operators, $A$ and $B$, do not *commute*, i.e.

$$AB \neq BA.$$

Recall that in quantum mechanics measurable quantities correspond to self adjoint operators. Related to this notion, whether or not two self adjoint linear operators commute has important physical consequences, as we will see. First, we consider an example that is particularly relevant to quantum mechanics.

*Example.* Recall the one dimensional momentum and position operators:

$$P = \frac{\hbar}{i}\frac{d}{dx}, \quad X.$$

Then we have:

$$
\begin{aligned}
(PX\psi)(x) &= \frac{\hbar}{i}\frac{d}{dx}(x\psi(x)), \\
&= \frac{\hbar}{i}\psi(x) + \frac{\hbar}{i}x\frac{d}{dx}\psi(x), \\
&= \frac{\hbar}{i}\psi(x) + (XP\psi)(x),
\end{aligned}
\tag{3.36}
$$

or

$$PX = \frac{\hbar}{i} + XP. \tag{3.37}$$

To be more precise, we have the following definition.

**Definition 26 (Commutator).** The commutator of two linear operators $A$ and $B$ is defined as:

$$[A, B] = AB - BA. \tag{3.38}$$

With this notation (3.37) becomes:

$$[X, P] = i\hbar. \tag{3.39}$$

Using the same idea, it is not hard to show that the following commutation relations hold in three dimensions.

$$[P_i, P_j] = 0, \quad [X_i, X_j] = 0, \quad [X_i, P_j] = i\hbar\delta_{ij}. \tag{3.40}$$

Commutators of linear operators satisfy a number of useful identities that we list below.

(i) $[A, B] = -[B, A]$
(ii) $[A, B]$ is linear in both $A$ and $B$.
(iii) $[A, BC] = B[A, C] + [A, B]C$.
(iv) $[A, [B, C]] + [B, [C, A]] + [C, [A, B]] = 0$

The proofs of these identities will be left as exercises.

### *Simultaneous Measurability*

Suppose $A$ and $B$ are self adjoint operators representing "observables", i.e. quantities that can be measured. Recall that the outcome of a measurement of $A$ (or $B$) can only be known with certainty if it is in an eigenstate. We are concerned here with the possibility of measuring $A$ and $B$ in the same basis. The following result is fundamental, and it highlights the significance of the commutator.

**Theorem 8.** *Suppose $A$ and $B$ are self adjoint operators on a finite dimensional, complex inner product space. Then $A$ and $B$ have a joint orthonormal basis of eigenvectors if and only if $[A, B] = 0$.*

**Proof.** First, let is suppose that

$$|n\rangle, \; n = 1, 2, 3, \ldots,$$

is an orthonormal basis of eigenvectors *for both* $A$ and $B$, i.e.,

$$A|n\rangle = a_n|n\rangle, \tag{3.41}$$

$$B|n\rangle = b_n|n\rangle. \tag{3.42}$$

A general state can be represented as follows:

$$|\psi\rangle = \sum_n c_n|n\rangle.$$

Then we have:

$$AB|\psi\rangle = \sum_n ABc_n|n\rangle$$

$$= \sum_n c_n AB|n\rangle,$$

$$= \sum_n c_n Ab_n|n\rangle,$$

$$= \sum_n c_n b_n A|n\rangle,$$

$$= \sum_n c_n b_n a_n|n\rangle,$$

$$= \sum_n c_n a_n b_n|n\rangle,$$

$$= \sum_n c_n a_n B|n\rangle,$$

$$= \sum_n c_n B a_n |n\rangle,$$

$$= \sum_n c_n B A |n\rangle,$$

$$= \sum_n B A c_n |n\rangle,$$

$$= B A |\psi\rangle. \tag{3.43}$$

Since the state $|\psi\rangle$ was completely arbitrary, this implies that

$$AB = BA.$$

Now let us suppose that $AB = BA$ and that $|n\rangle$, $n = 1, 2, 3, \ldots$ is an orthonormal basis for $A$ (i.e. $A|n\rangle = a_n|n\rangle$). We will show that this implies that $|n\rangle$, $n = 1, 2, 3, \ldots$ is an orthonormal basis for $B$.

Now since $AB = BA$ we have:

$$AB|n\rangle = BA|n\rangle = a_n B|n\rangle. \tag{3.44}$$

This calculation implies that $B|n\rangle$ is also an eigenvector of $A$ with eigenvalue $a_n$. There are two cases to consider.

$\boxed{a_n \text{ is non-degenerate.}}$ In this case there is only one eigenvector corresponding to the eigenvalue $a_n$. Therefore since $B|n\rangle$ is also an eigenvector of $A$, it must be proportional to $|n\rangle$, i.e.

$$B|n\rangle = b_n|n\rangle.$$

In other words, $|n\rangle$ is also an eigenvector of $B$. Hence, $|n\rangle, n = 1, 2, 3, \ldots$ are also eigenvectors for $B$.

Now we consider the other case.

$\boxed{a_n \text{ is degenerate.}}$ Suppose $|n_j\rangle$, $j = 1, \ldots, d$ are all the orthonormal eigenvectors of $A$ corresponding to the eigenvalue $a_n$. We know from (3.44) that $B|n\rangle$ is an eigenvector of $A$, with eigenvalue $a_n$, but it is not necessarily an eigenvector of $B$ since it need not be proportional to $|n\rangle$

Now let

$$W = \text{span} \left\{ |n_j\rangle, \; j = 1, \ldots, d \right\}.$$

You should be able to verify to yourself that $W$ is an invariant subspace for $A$, i.e. $A : W \to W$. Now you need to go a step further and, using (3.44) and argue that $W$ is an invariant subspace for $B$, i.e. $B : W \to W$.

In this case, it makes sense to restrict $B$ to $W$. Now we want to argue that $B$ restricted to $W$ is self-adjoint. This can be seen as follows. Express $B$ as follows:

$$B|n_j\rangle = \sum_{i=1}^{d} |n_i\rangle\langle n_i|B|n_j\rangle.$$

Then the matrix elements of $B$ satisfy the following:

$$B_{ij} = \langle n_i|B|n_j\rangle = \overline{\langle n_j|B^\dagger|n_i\rangle} = \overline{\langle n_j|B|n_i\rangle} = \overline{B_{ji}}.$$

Hence, $B$, restricted to $W$, is self adjoint. Therefore, $B\big|_W$ can be diagonalised, i.e. we can find $d$ orthonormal eigenvectors, $|\tilde{n}_j\rangle$, $j = 1, \ldots, d$, and $B|\tilde{n}_j\rangle = b_{\tilde{n}_j}|\tilde{n}_j\rangle$. Now it follows from (3.44) that

$$A|\tilde{n}_j\rangle = a_{\tilde{n}_j}|\tilde{n}_j\rangle.$$

In this way, we are able to find a set of orthonormal eigenvectors for both $A$ and $B$, restricted to $W$. This procedure can be repeated for every degenerate eigenvalue, if necessary, to construct a set of orthonormal eigenvectors for $A$ and $B$ on the entire space. $\qquad\square$

(Mathematically, this proof is a bit "sloppy". See if you can figure out where it should be "cleaned up".)

This is an important result since it often turns out that in order to specify the state of a system, we need to measure more than one quantity, i.e. we require more than one observable. We will see a concrete example of this when we study angular momentum. For the observables to be "compatible" we need to be able to express them in a common basis, i.e. the observables must commute.

Finally, we have not been explicit about how this result leads to simultaneous measurability. Recall we have denoted $|n\rangle$ to be the orthonormal eigenbasis for both $A$ and $B$ with corresponding eigenvalues $a_n$ and $b_n$, respectively. Consider an arbitrary state $|\psi\rangle$, and we measure $A$ in the state $|\psi\rangle$ and obtain the eigenvalue $a_n$. Therefore we know that after this measurement the system is in the state $|n\rangle$. Therefore since $|n\rangle$ is also an eigenstate of $B$, measurement of $B$ yields $b_n$. In this way the existence of a common orthonormal basis for the self-adjoint operators $A$ and $B$ yields simultaneous measurability.

## 3.3  Uncertainty Relations

In this section we will derive the general form of the uncertainty principle due to H. P. Robertson for self-adjoint operators.[1] As a preamble to a discussion of the uncertainty principle in quantum mechanics it is difficult to do much better than the description that Robertson gave in his original paper, from which we quote:

> The uncertainty principle is one of the most characteristic and important consequences of the new quantum mechanics. This principle, as formulated by Heisenberg for two conjugate quantum-mechanical variables, states that the accuracy with which two such variables can be measured simultaneously is subject to the restriction that the product of the uncertainties in the two measurements is at least of order $h$ (Planck's constant).

Prior to Robertson's work, the uncertainty principle began to be established in quantum theory through the work of Heisenberg,[2] Kennard,[3] and Weyl.[4] Heisenberg described[5] a very insightful and intriguing thought experiment (Heisenberg's microscope) that is well-worth considering in order to grasp the deeper physical aspects of the uncertainly principle. A description of the uncertainty principle from a contemporary point of view is given in the paper of Furuta,[6] which also contains a description of its historical development.

We now give the proof of Robertson's mathematical formulation of the uncertainty principle. Recall that the dispersion of $A$ in the state $|\psi\rangle$ is

---

[1] H. P. Robertson. The uncertainty principle. *Physical Review*, 34(1):163, 1929.

[2] W. Heisenberg. Über den anschaulichen inhalt der quantentheoretischen kinematik und mechanik. In *Original Scientific Papers Wissenschaftliche Originalarbeiten*, pages 478–504. Springer, 1985.

[3] E. H. Kennard. Zur quantenmechanik einfacher bewegungstypen. *Zeitschrift für Physik*, 44(4-5):326–352, 1927.

[4] H. Weyl. Gruppentheorie und quantenmechanik, hirzel, leipzig. *Theory of Groups and Quantum Mechanics, 2nd ed.(1931), transl. H. P. Robertson, Dover, NY (1950)*, pages 100–101, 1928.

[5] W. Heisenberg. *The physical principles of the quantum theory*. Courier Corporation, 1949.

[6] A. Furuta. One thing is certain: Heisenberg's uncertainty principle is not dead. *Scientific American*, 2012.

defined as:

$$\Delta_\psi(A) = \left[E_\psi(A^2) - E_\psi(A)^2\right]^{\frac{1}{2}}, \tag{3.45}$$

where

$$E_\psi(A) = \langle\psi|A|\psi\rangle, \qquad \langle\psi|\psi\rangle = 1.$$

We now derive the following "generalised" uncertainty relation.

**Theorem 9.** *Let $A$ and $B$ be self-adjoint operators, then*

$$\Delta_\psi(A)\Delta_\psi(B) \geq \frac{1}{2}|E_\psi([A, B])|. \tag{3.46}$$

*We have equality if*

$$\{(A - E_\psi(A)) - it(B - E_\psi(B))\}|\psi\rangle = 0. \tag{3.47}$$

*for some $t \in \mathbb{R}$.*

**Proof.** Define

$$[A, B] = iC.$$

Then it is a simple exercise to show that

$$C^\dagger = C.$$

We will use this result in the course of the proof.

Now consider

$$(A - itB), \quad t \in \mathbb{R}.$$

Then

$$(A - itB)^\dagger = A + itB, \quad \text{(note that we used the fact that $t$ is real here.)}$$

and therefore:

$$\begin{aligned}
(A - itB)^\dagger(A - itB) &= (A + itB)(A - itB), \\
&= A^2 - it(AB - BA) + t^2 B^2, \\
&= A^2 + tC + t^2 B^2. \tag{3.48}
\end{aligned}$$

Using this result, it follows that:

$$\begin{aligned}
E_\psi(A^2) + tE_\psi(C) + t^2 E_\psi(B^2) &= \langle\psi|(A - itB)^\dagger(A - itB)|\psi\rangle, \\
&= \|(A - itB)|\psi\rangle\|^2, \\
&\geq 0, \tag{3.49}
\end{aligned}$$

and this expression is zero if and only if:

$$(A - itB)|\psi\rangle = 0. \tag{3.50}$$

Now we choose $t$ such that the left hand side of (3.49) is minimal. The choice of $t$ that accomplishes this is obtained by differentiating the left hand side of (3.49) with respect to $t$ and setting the result to zero:

$$\frac{d}{dt}\left(E_\psi(A^2) + tE_\psi(C) + t^2 E_\psi(B^2)\right) = E_\psi(C) + 2tE_\psi(B^2) = 0,$$

$$\Rightarrow \quad t = -\frac{E_\psi(C)}{2E_\psi(B^2)}.$$

Substituting this value of $t$ into the left hand side of (3.49) gives:

$$E_\psi(A^2) - \frac{E_\psi(C)^2}{2E_\psi(B^2)} + \frac{E_\psi(C)^2}{4E_\psi(B^2)} \geq 0,$$

which simplifies to

$$E_\psi(A^2) - \frac{1}{4}\frac{E_\psi(C)^2}{E_\psi(B^2)} \geq 0,$$

or

$$E_\psi(A^2)E_\psi(B^2) \geq \frac{1}{4}E_\psi(C)^2. \tag{3.51}$$

Now (3.51) is valid for any self-adjoint operators, $A$ and $B$. We therefore make the following substitution into (3.51):

$$A \to A - E_\psi(A), \quad B \to B - E_\psi(B). \tag{3.52}$$

It then follows from straightforward calculations that:

$$E_\psi\left(A - E_\psi(A)\right) = E_\psi(A^2) - E_\psi(A)^2 = \Delta_\psi(A)^2, \tag{3.53}$$

$$E_\psi\left(B - E_\psi(B)\right) = E_\psi(B^2) - E_\psi(B)^2 = \Delta_\psi(B)^2, \tag{3.54}$$

and

$$[A - E_\psi(A), \, B - E_\psi(B)] = [A, B] = iC. \tag{3.55}$$

Substituting (3.53) and (3.54) into the left side of (3.51), and applying (3.55) to the right side of (3.51) gives:

$$\Delta_\psi(A)^2 \, \Delta_\psi(B)^2 \geq \frac{1}{4}E_\psi\left(\frac{1}{i}[A, B]\right)^2. \tag{3.56}$$

Then taking the square root of the result gives:

$$\Delta_\psi(A)\,\Delta_\psi(B) \geq \left|\frac{1}{2i}E_\psi([A,B])\right| = \frac{1}{2}\left|E_\psi([A,B])\right|. \qquad (3.57)$$

Finally, note that (3.47) is obtained by substituting (3.52) into (3.50).  $\square$

It is important to realize that, strictly speaking, this is purely a mathematical result describing statistical properties of self-adjoint operators. The physics is introduced when the self-adjoint operators correspond to physical observables, and we next consider examples of this

*Generalized Uncertainty Relation for the Position and Momentum Operators.* We now compute the generalised uncertainty relation for the position and momentum operators;

$$\Delta_\psi(P)\Delta_\psi(X) \geq \frac{1}{2}\left|E_\psi[P,X]\right|.$$

Recall that:

$$[P,X] = -i\hbar,$$

and therefore

$$\left|E_\psi\left([P,X]\right)\right| = \hbar.$$

So we have:

$$\Delta_\psi(P)\Delta_\psi(X) \geq \frac{\hbar}{2}.$$

Recall that, in general, the dispersion of an observable depends upon the state in which the measurement is made. However, the right hand side of this relation is independent of the state. So this uncertainty relation is true *for any* (normalised) state $|\psi\rangle$ in which the measurements of $X$ and $P$ are made.

*A State of Minimal Uncertainty.* Recall that we have equality of the generalised uncertainty relation (3.46), provided (3.47) is satisfied. In other words, substituting (3.52) into (3.49) gives the following criterion for minimizing the uncertainty:

$$\{A - E(A) - it(B - E(B))\}\,|\psi\rangle = 0. \qquad (3.58)$$

Now we consider a specific example. We consider the (one dimensional) position and momentum operators, and we want to find the state, $|\psi\rangle$ for which the uncertainty is minimal. In particular, we have:

$$A = P = \frac{\hbar}{i}\frac{d}{dx}, \; B = X, \; \tilde{p} \equiv E_\psi(P), \; \tilde{x} \equiv E_\psi(X).$$

With these substitutions (3.58) becomes:

$$\left(\frac{\hbar}{i}\frac{d}{dx} - \tilde{p}\right)\psi = it(x - \tilde{x})\psi \tag{3.59}$$

or

$$\frac{d}{dx}\psi = \frac{i}{\hbar}\left(\tilde{p} + it(x - \tilde{x})\right)\psi. \tag{3.60}$$

This equation can be solved to give:

$$\psi(x) = c\,\exp\left(\frac{i}{\hbar}\tilde{p}x - \frac{t}{2\hbar}(x - \tilde{x})^2\right), \tag{3.61}$$

and the constant $c$ can be chosen to satisfy normalization. You have seen this particular function when we discussed wavepackets Section 2.3. It is referred to as a "Gaussian", and this is another example of how wavefunctions of this form play an important role in quantum mechanics. We remark that the calculations in Section 2.3 are useful in explaining how to go from (3.60) to (3.61).

## 3.4  The Harmonic Oscillator

The (unforced, undamped) harmonic oscillator is a basic, "paradigm" dynamical system that is essential to understand.

The Hamitonian for the classical harmonic oscillator is given by:

$$H = \frac{1}{2m}p^2 + \frac{1}{2}m\omega^2 x^2, \tag{3.62}$$

where $m$ denotes the mass of the (point) particle, $\omega$ is the frequency of oscillation, and $p$ and $x$ are the coordinates corresponding the the momentum and position, respectively. Hamilton's differential equations are given by:

$$\dot{x} = \frac{\partial H}{\partial p} = \frac{p}{m},$$

$$\dot{p} = -\frac{\partial H}{\partial x} = -m\omega^2 x, \tag{3.63}$$

and "solving" the classical harmonic oscillator might mean finding the solutions to Hamilton's differential equations.

We now turn our attention to the quantum mechanical harmonic oscillator. The quantum mechanical Hamiltonian is obtain from the classical Hamiltonian by replacing $p$ and $x$ with the corresponding operators:

$$H = \frac{1}{2m}P^2 + \frac{1}{2}m\omega^2 X^2. \tag{3.64}$$

For us, "solving" the quantum mechanical oscillator equation will mean finding the eigenvalues and eigenstates of the time-independent Schrödinger equation:

$$H|\psi\rangle = E|\psi\rangle, \tag{3.65}$$

where $E$ is the eigenvalue and $H$ is given by (3.64). It will be useful to recall the commutation relation between $X$ and $P$, which we re-write here:

$$[X, P] = i\hbar. \tag{3.66}$$

We will introduce a method that is algebraic in nature, and heavily uses the commutation relation (3.66) (we will revisit the same method when we study quantum mechanical angular momentum). It goes by the names of "raising and lowering operators". "creation and annihilation operators", "ladder operators" or the method of factorization. The different names are used in the different areas in which it is used. It is surprising that it is not taught as a general mathematical approach in ordinary differential equations and partial differential equations courses. More details on the method can be found in this reference.[7]

We begin by introducing the operators:

$$a = P - im\omega X,$$

$$a^\dagger = P + im\omega X. \tag{3.67}$$

It is straightforward to verify that they satisfy the following commutation relation:

$$[a, a^\dagger] = [P - im\omega X, P + im\omega X],$$

$$= im\omega[P, X] - im\omega[X, P],$$

$$= 2m\omega\hbar. \tag{3.68}$$

---

[7]B. Mielnik and O. Rosas-Ortiz. Factorization: little or great algorithm? *Journal of Physics A: Mathematical and General*, 37(43):10007, 2004.

Moreover, it is easy to verify that:

$$a^\dagger a = (P + im\omega X)(P - im\omega X),$$
$$= P^2 + im\omega(XP - PX) + m^2\omega^2 X^2,$$
$$= 2m\left(H - \frac{\omega\hbar}{2}\right), \tag{3.69}$$

and

$$aa^\dagger = [a, a^\dagger] + a^\dagger a,$$
$$= 2m\left(H + \frac{\omega\hbar}{2}\right). \tag{3.70}$$

We define the *number operator* as follows:

$$N = \frac{1}{2m\hbar\omega} a^\dagger a. \tag{3.71}$$

Note that using (3.69) and (3.71), the Hamiltonian, (3.64), can be expressed in terms of the number operator as follows:

$$H = \hbar\omega\left(N + \frac{1}{2}\right). \tag{3.72}$$

Below we collect together some useful properties of the number operator.

*Properties of the number operator.*

- $N$ is self adjoint. This follows from the calculation:

$$N^\dagger = \frac{a^\dagger a}{2m\hbar\omega} = N. \tag{3.73}$$

- $N$ is positive. This follows from the calculation:

$$\langle\psi|N|\psi\rangle = \langle\psi|\frac{a^\dagger a}{2m\hbar\omega}|\psi\rangle = \frac{1}{2m\hbar\omega}(a|\psi\rangle, a|\psi\rangle) \geq 0. \tag{3.74}$$

- Commutators:

$$[N, a] = \frac{1}{2m\hbar\omega}[a^\dagger a, a] = \frac{1}{2m\hbar\omega}\left\{a^\dagger[a, a] + [a^\dagger, a]a\right\} = -a, \tag{3.75}$$

$$[N, a^\dagger] = \frac{1}{2m\hbar\omega}[a^\dagger a, a^\dagger] = \frac{1}{2m\hbar\omega}\left\{a^\dagger[a, a^\dagger] + [a^\dagger, a^\dagger]a\right\} = a^\dagger. \tag{3.76}$$

Now consider the general equation for the eigenvalues and eigenvectors of $N$:[8]

$$N|\eta\rangle = \eta|\eta\rangle. \tag{3.77}$$

Our goal is to determine the eigenvalues $\eta$ and the eigenstates $|\eta\rangle$ of $N$ by using the following properties:

$$\langle\eta|N|\eta\rangle \geq 0, \qquad \eta = \frac{\langle\eta|N|\eta\rangle}{\langle\eta|\eta\rangle} \geq 0. \tag{3.78}$$

It follows from (3.72) that the eigenvalues and eigenstates of the number operator are also eigenvalues and eigenstates of the Hamiltonian of the harmonic oscillator.

$$\boxed{a^\dagger|\eta\rangle \text{ is an eigenstate of } N \text{ with eigenvalue } \eta + 1.}$$

Using (3.76) we have:

$$\begin{aligned}
Na^\dagger|\eta\rangle &= \left(a^\dagger N + a^\dagger\right)|\eta\rangle, \\
&= a^\dagger N|\eta\rangle + a^\dagger|\eta\rangle, \\
&= (\eta + 1)a^\dagger|\eta\rangle,
\end{aligned} \tag{3.79}$$

which implies that $a^\dagger|\eta\rangle$ is an eigenstate of $N$ with eigenvalue $\eta+1$, *provided* $a^\dagger|\eta\rangle \neq 0$, which we show is the case.

$$\boxed{\textbf{Technical Point: } a^\dagger|\eta\rangle \neq 0.}$$

$$\begin{aligned}
\left(a^\dagger|\eta\rangle, a^\dagger|\eta\rangle\right) &= \langle\eta|aa^\dagger|\eta\rangle = \langle\eta|a^\dagger a + 2m\hbar\omega|\eta\rangle, \\
&= \langle\eta|(2m\hbar\omega)N + 2m\hbar\omega|\eta\rangle, \\
&= 2m\hbar\omega\langle\eta|N|\eta\rangle + 2m\hbar\omega\langle\eta|\eta\rangle, \\
&= 2m\hbar\omega\eta\langle\eta|\eta\rangle + 2m\hbar\omega\langle\eta|\eta\rangle, \\
&= 2m\hbar\omega(\eta + 1)\langle\eta|\eta\rangle, \\
&> 0,
\end{aligned} \tag{3.80}$$

from which it follows that:

$$a^\dagger|\eta\rangle \neq 0, \tag{3.81}$$

$$\boxed{a|\eta\rangle \text{ is an eigenstate of } N \text{ with eigenvalue } \eta - 1.}$$

---

[8]Note that when we write down such general equations, we are *assuming* that $|\eta\rangle \neq 0$, since zero eigenvectors are not interesting. If we give an argument for the existence of a general eigenvector, we then need to prove afterward (if not done during the course of the proof) that the eigenvector is not zero.

Using (3.75), we have:

$$Na|\eta\rangle = (aN - a)|\eta\rangle,$$
$$= (\eta - 1)a|\eta\rangle, \tag{3.82}$$

which implies that $a|\eta\rangle$ is an eigenstate of $N$ with eigenvalue $\eta - 1$, *provided* $a|\eta\rangle \neq 0$, which we show is the case with the following calculation:

---
**Technical Point:** $a|\eta\rangle \neq 0.$
---

$$(a|\eta\rangle, a|\eta\rangle) = \langle\eta|a^\dagger a|\eta\rangle = 2m\hbar\omega\langle\eta|N|\eta\rangle = 2m\hbar\omega\eta\langle\eta|\eta\rangle \geq 0, \tag{3.83}$$

from which it follows that $a|\eta\rangle = 0$ if and only if $\eta = 0$ (since we are assuming that $|\eta\rangle \neq 0$).

From these calculations we see that it is natural to refer to $a^\dagger$ as the *raising operator* and $a$ as the *lowering operator*.

---
$N$ has a smallest eigenvalue, which is zero.
---

Since $N \geq 0$, it has a smallest eigenvalue, $\eta_{min}$. We now argue that $\eta_{min} = 0$. Suppose

$$a|\eta_{min}\rangle \neq 0. \tag{3.84}$$

Then we have:

$$Na|\eta_{min}\rangle = (\eta_{min} - 1)\, a|\eta_{min}\rangle. \tag{3.85}$$

This implies that $a|\eta_{min}\rangle$ is an eigenstate of $N$ with eigenvalue $\eta_{min} - 1$, which contradicts the fact that $\eta_{min}$ is the minimum eigenvalue of $N$. This contradiction means that the assumption (3.84) is false. Hence we have:

$$a|\eta_{min}\rangle = 0. \tag{3.86}$$

From this calculation, and the conclusion of (3.83), it follows that:

$$\eta_{min} = 0. \tag{3.87}$$

We summarise our results.

- $N$ has eigenvalues $0, 1, 2, 3, \ldots$.
- It follows from (3.79) that $a^\dagger|n\rangle$ is proportional to $|n+1\rangle$, and the proportionality constant is chosen so that the eigenstates are normalised
- It follows from (3.83) that $a|n\rangle$ is proportional to $|n-1\rangle$, and the proportionality constant is chosen so that the eigenstates are normalised.

The eigenstate with smallest eigenvalue, $|0\rangle$ is referred to as the *ground state* or *vacuum state*, and we normalise this eigenstate such that:

$$\langle 0|0 \rangle = 1. \tag{3.88}$$

All other eigenstates can be obtained from $|0\rangle$ by acting on it with the raising operator, $a^\dagger$:

$$|n\rangle = c_n (a^\dagger)^n |0\rangle. \tag{3.89}$$

It follows from a calculation that requiring

$$\langle n|n \rangle = 1,$$

implies that

$$c_n^2 = \frac{1}{\langle 0|a^n (a^\dagger)^n |0\rangle}, \tag{3.90}$$

or

$$c_n = \frac{1}{\sqrt{n!}\,(2m\hbar\omega)^{\frac{n}{2}}}. \tag{3.91}$$

The calculation that leads to this normalisation factor are described in Problem 5 at the end of this chapter.

Now recall the form of the Hamiltonian for the harmonic oscillator expressed in terms of the number operator given in (3.72). Hence, we have shown that:

$$H|n\rangle = \hbar\omega \left( n + \frac{1}{2} \right) |n\rangle \qquad n = 0, 1, 2, \ldots \tag{3.92}$$

In other words, we are able to determine the eigenvalues of $H$ from the properties of the raising and lowering operators (in particular, from the number operator). Now we turn our attention to finding the eigenstates explicitly.

First, we determine $|0\rangle$. From (3.86) we have:

$$a|0\rangle = 0. \tag{3.93}$$

From the form of $a$ given in (3.67), (3.93) has the form:

$$(P - im\omega X)\,|0\rangle = 0, \tag{3.94}$$

or, substituting into this equation the explicit expressions for the position and momentum operators:

$$\frac{\hbar}{i}\psi_0'(x) - im\omega x\psi_0(x) = 0, \tag{3.95}$$

or

$$\psi_0'(x) = -\frac{m\omega}{\hbar}x\psi_0(x). \tag{3.96}$$

This ordinary differential equation can be solved explicitly to give:

$$\psi_0(x) = c\exp\left(-\frac{m\omega}{2\hbar}x^2\right). \tag{3.97}$$

The constant $c$ is evaluated through the normalisation condition:

$$1 = \int_{-\infty}^{+\infty} |\psi_0(x)|^2 dx = \int_{-\infty}^{+\infty} c^2 \exp\left(-\frac{m\omega}{\hbar}x^2\right) dx = c^2\sqrt{\frac{\pi\hbar}{m\omega}}, \tag{3.98}$$

which gives:

$$c = \left(\frac{m\omega}{\pi\hbar}\right)^{\frac{1}{4}}. \tag{3.99}$$

With $\psi_0(x)$ determined, the remaining eigenstates can be computed by acting on $\psi_0(x)$ with the raising operator:

$$|n\rangle = c_n\left(a^\dagger\right)^n |0\rangle, \quad c_n = \frac{1}{\sqrt{n!}(2m\hbar\omega)^{\frac{n}{2}}}. \tag{3.100}$$

or, more explicitly,

$$\psi_n(x) = c_n\left(\frac{\hbar}{i}\frac{d}{dx} + im\omega x\right)^n \psi_0(x). \tag{3.101}$$

## Problems

1. *General properties of commutators.* Check, using the definition of the commutator, that

   (a) $[A, B] = -[B, A]$,
   (b) $[A, B]$ is linear in both $A$ and $B$,
   (c) $[A, BC] = B[A, C] + [A, B]C$,
   (d) $[A, [B, C]] + [B, [C, A]] + [C, [A, B]] = 0$,

   for operators $A$, $B$ and $C$.

2. *Canonical commutation relations.*

   (a) Prove by induction that

   $$[X^n, P] = i\hbar n X^{n-1}$$

   and that

   $$[X, P^n] = i\hbar n P^{n-1}$$

   where $n$ is a positive integer.

(b) Using these results show that if $f(x)$ can be expanded in a polynomial in $x$ and $g(p)$ can be expanded in a polynomial in $p$, then

$$[f(X), P] = i\hbar f'(X)$$

and

$$[X, g(P)] = i\hbar g'(P)$$

This result is derived for any differentiable $f$ and $g$ in Hannabuss, problem 7.4.

(c) More generally, prove by induction, that if two operators, $A$ and $B$, commute with their commutator (i.e. $[A, [A, B]] = [B, [A, B]] = 0$) then

$$[A, B^n] = nB^{n-1}[A, B] \quad \text{and} \quad [B, A^n] = -nA^{n-1}[A, B]$$

3. *Commutators again.* Consider two operators $N$ and $A$ which satisfy $[N, A] = -A$. And define the operator $C(\theta)$ depending on a real parameter $\theta$ by

$$C(\theta) = e^{\theta N} A e^{-\theta N}$$

(a) Using the definition of the exponential of an operator, show that $\frac{d}{d\theta}(e^{\theta N}) = N e^{\theta N}$.

(b) Show that $\frac{dC(\theta)}{d\theta} = -C(\theta)$.

(c) Assuming that the differential equation in part (b) has a unique solution given $C(0)$, show that $C(\theta) = e^{-\theta} A$.

4. *Creation and annihilation operators, 1.* Let $X$ and $P$ be the standard position and momentum observables, satisfying $[X, P] = i\hbar$ ($\hbar$ is Planck's constant). The creation and annihilation operators $a^\dagger$ and $a$ are defined by

$$a^\dagger = P + im\omega X; \quad a = P - im\omega X$$

where $m$ and $\omega$ are constants.

(a) Show that $[a, a^\dagger] = 2m\omega\hbar.$

(b) Show that

$$a(a^\dagger)^n = (a^\dagger)^n a + n2m\omega\hbar(a^\dagger)^{(n-1)},$$

where $n$ is a non-negative integer.

(c) Show also that

$$ae^{\lambda a^\dagger} = e^{\lambda a^\dagger}(a + \lambda 2m\omega\hbar)$$

where $\lambda$ is a complex constant and $e^{\lambda a^\dagger}$ is defined by

$$e^{\lambda a^\dagger} = \sum_{n=0}^{\infty} \lambda^n (a^\dagger)^n / n!.$$

5. *Creation and annihilation operators, 2.* Let $a$ and $a^\dagger$ be the operators defined in the previous question, and let $|0\rangle$ be the normalised vacuum state defined by

$$a\,|0\rangle = 0.$$

The normalised state $|n\rangle$ is defined by

$$|n\rangle = c_n (a^\dagger)^n \, |0\rangle,$$

where $c_n$ is a positive constant, and $n$ is a non-negative integer.

(a) Show that $2m\omega\hbar n c_n^2 = c_{n-1}^2$, and hence, or otherwise, calculate $c_n$.
(b) Using part (b) of the previous question, show that $a\,|n\rangle = d_n\,|n-1\rangle$, where $d_n$ is a constant, and calculate $d_n$.
(c) Show that $\langle n_1\,|n_2\rangle = 0$ if $n_1 \neq n_2$.
(d) By writing $X$ and $P$ in terms of $a$ and $a^\dagger$, calculate $\langle n_1|\,X\,|n_2\rangle$ and $\langle n_1|\,P\,|n_2\rangle$.

6. Prove that the dispersion of an operator $A$ in the state $|\psi\rangle$ vanishes *if and only if* $|\psi\rangle$ is an eigenvector of $A$.

7. On a space spanned by the orthonormal basis $|1\rangle, |2\rangle, |3\rangle$, let $G$ be the operator defined by

$$G\,|1\rangle = \frac{5}{4}\,|1\rangle + \frac{\sqrt{3}}{4}\,|2\rangle$$

$$G\,|2\rangle = \frac{\sqrt{3}}{4}\,|1\rangle + \frac{7}{4}\,|2\rangle$$

$$G\,|3\rangle = 2\,|3\rangle$$

(a) Find the eigenvalues of $G$ and eigenvectors of $G$.
(b) Write $G$ in a spectral decomposition $G = \lambda_1 P_1 + \lambda_2 P_2$, where one of the projection operators is two-dimensional; you should give $P_1$ and $P_2$ explicitly in Dirac notation.

(c) Let a particle initially be in the state $|\phi\rangle = \cos\theta\,|1\rangle + \sin\theta\,|3\rangle$, where $\theta$ is a constant. Find the probability that the outcome, $+2$, is found when $G$ is measured, and find the state immediately after measurement.

(d) Find the expected value of $G$ in the state $|\phi\rangle$ and the dispersion of $G$ in this state.

(e) Is there a value of $\theta$ for which the dispersion is zero? Comment on this.

8. Let $H$ denote a self-adjoint Hamiltonian operator and $|\psi\rangle$ an eigenstate of $H$, i.e.

$$H|\psi\rangle = E|\psi\rangle.$$

For any other operator $A$, $\langle A\rangle \equiv \langle\psi|A|\psi\rangle$ denotes the expectation value of $A$ in the state $|\psi\rangle$.

(a) For any operator $A$ show that

$$\langle\psi|[H, A]|\psi\rangle = 0. \tag{3.102}$$

(b) If $X$ and $P$ denote the position and momentum operators, respectively, show that:

    (i) $[P^2, X] = -2i\hbar P$,
    (ii) $[P^2, XP] = -2i\hbar P^2$,
    (iii) $[X^N, XP] = Ni\hbar X^N$, where $N$ is an integer, $N \geq 1$.

Now suppose we consider a self-adjoint Hamiltonian operator of the form:

$$H = T + V = \frac{P^2}{2m} + kX^N, \tag{3.103}$$

where $X$ and $P$ denote the position and momentum operators, respectively, and $N$ is an integer, $N \geq 1$, $m > 0$, and $k$ is a real number.

(c) Using (3.102) and $A = X$, show that $\langle P\rangle = 0$.

(d) Using (3.102) and $A = XP$, show that $2\langle T\rangle = N\langle V\rangle$.

(e) Compute the variance, $\langle(\Delta P)^2\rangle = \langle P^2\rangle - \langle P\rangle^2$, of the momentum operator $P$ in the state $|\psi\rangle$ and express it in terms of the expectation value of the kinetic energy, $T$, in the state $|\psi\rangle$.

9. Consider the state space $\mathbb{C}^2$ with orthonormal basis,

$$|1\rangle = \begin{pmatrix} 1 \\ 0 \end{pmatrix}, \qquad |2\rangle = \begin{pmatrix} 0 \\ 1 \end{pmatrix}.$$

We define the operator $A$ on $\mathbb{C}^2$ with respect to this basis as follows:

$$A|1\rangle = 2|1\rangle - i|2\rangle,$$

$$A|2\rangle = i|1\rangle + 2|2\rangle.$$

(a) Write down the matrix representations of $A$ with respect to the basis $|1\rangle$, $|2\rangle$.

(b) Show that $A$ is self-adjoint.

(c) Show that the eigenvalues of $A$ are 1 and 3. Let $|e_1\rangle$ denote the normalized eigenstate corresponding to the eigenvalue 1 and let $|e_2\rangle$ denote the normalized eigenstate corresponding to the eigenvalue 3. Compute expressions for $|e_1\rangle$ and $|e_2\rangle$.

(d) Express the basis vectors $|1\rangle$ and $|2\rangle$ in terms of $|e_1\rangle$ and $|e_2\rangle$.

(e) Show that $|e_1\rangle$ and $|e_2\rangle$ are orthonormal.

(f) Express $A$ in the basis $\{|1\rangle, |2\rangle\}$ using Dirac notation.

(g) Express $A$ in the basis $\{|e_1\rangle, |e_2\rangle\}$ using Dirac notation.

(h) Suppose the system is in the state $|1\rangle$ and we wish to measure A in this state. Compute the probability of measuring 1 in the state $|1\rangle$, $\mathrm{Pr}_{|1\rangle}(1)$.

(i) Suppose the system is in the state $|1\rangle$ and we wish to measure A in this state. Compute the probability of measuring 3 in the state $|1\rangle$, $\mathrm{Pr}_{|1\rangle}(3)$.

(j) Suppose the system is in the state $|2\rangle$ and we wish to measure A in this state. Compute the probability of measuring 1 in the state $|2\rangle$, $\mathrm{Pr}_{|2\rangle}(1)$.

(k) Suppose the system is in the state $|2\rangle$ and we wish to measure A in this state. Compute the probability of measuring 3 in the state $|2\rangle$, $\mathrm{Pr}_{|2\rangle}(3)$.

(l) If the result of the measurement of $A$ in the state $|1\rangle$ is 1, what is the state of the system after measurement?

(m) If the result of the measurement of $A$ in the state $|1\rangle$ is 3, what is the state of the system after measurement?

(n) If the result of the measurement of $A$ in the state $|2\rangle$ is 1, what is the state of the system after measurement?

(o) If the result of the measurement of $A$ in the state $|2\rangle$ is 3, what is the state of the system after measurement?

10. (The Inverted Harmonic Oscillator) Consider a quantum particle of constant mass $m$ moving in the one dimensional inverted harmonic oscillator potential. The Hamiltonian is given by

$$H = \frac{P^2}{2m} - \frac{1}{2}m\omega^2 X^2,$$

where $P$ and $X$ are the one dimensional momentum and position operators, respectively, and $m$ and $\omega$ are positive, real constants. We define

$$a = P + m\omega X,$$

$$b = P - m\omega X,$$

$$N = \frac{ba}{2m\hbar\omega}.$$

(a) Are $a$ and $b$ self-adjoint?
(b) Is $N$ self-adjoint?
(c) Calculate $[N, a]$.
(d) Calculate $[N, b]$.
(e) Express $H$ in terms of $N$.

11. Suppose a quantum particle in one dimension is described by the following wavefunction:

$$\psi(x) = \left(\frac{2a}{\pi}\right)^{\frac{1}{4}} e^{-ax^2}, \quad a > 0.$$

You may use the following two integrals.

$$\int_{-\infty}^{\infty} e^{-2ax^2}\, dx = \sqrt{\frac{\pi}{2a}},$$

$$\int_{-\infty}^{\infty} x^2 e^{-2ax^2}\, dx = \frac{1}{4a}\sqrt{\frac{\pi}{2a}}.$$

(a) Calculate $\langle x \rangle$.
(b) Calculate $\langle x^2 \rangle$.
(c) Calculate $\langle p \rangle$.
(d) Calculate $\langle p^2 \rangle$.
(e) Calculate $\Delta x\, \Delta p$.

12. Consider the motion of a quantum particle of constant mass $m$ on $\mathbb{R}$ described by the Hamiltonian:

$$H = cP,$$

where $P$ is the momentum operator and $c > 0$ is a real number.

(a) Let $\langle X \rangle_0 \equiv x_0$ and $\langle P \rangle_0 \equiv p_0$ denote the expectation values of $X$ and $P$, respectively, at $t = 0$ and let $\langle X \rangle_t$ and $\langle P \rangle_t$ denote the expectation values of $X$ and $P$, respectively, at time $t$. Write down Ehrenfest's equations and solve for $\langle X \rangle_t$ and $\langle P \rangle_t$. .

(b) Show that $\sigma_P^2(t) = \langle (P - \langle P \rangle)^2 \rangle$ is constant in time.

(c) Calculate $\frac{d}{dt}\langle X^2 \rangle$.

(d) Calculate $\langle X^2 \rangle$.

(e) Show that $\sigma_X^2(t) = \langle (X - \langle X \rangle)^2 \rangle$ is constant in time.

13. Let $H$ denote a Hamiltonian operator acting on a Hilbert space, $A$ denotes a linear operator on the Hilbert space and $|\psi\rangle$ denotes a (normalized) state vector in that Hilbert space. We denote the expectation value of $A$ in the state $|\psi\rangle$ as $\langle \psi | A | \psi \rangle \equiv \langle A \rangle$.

(a) Using Dirac notation, show that

$$\frac{d}{dt}\langle A \rangle = \frac{1}{i\hbar}\langle [A, H] \rangle + \left\langle \frac{\partial A}{\partial t} \right\rangle,$$

where $[A, H] = AH - HA$.

Parts (b) and (c) refer to the one dimensional Hamiltonian set-up that we now describe.

Consider a one dimensional Hamiltonian:

$$H(x, p) = \frac{p^2}{2m} + V(x),$$

where $V(x)$ has the form of a polynomial:

$$V(x) = a_0 + a_1 x + \cdots + a_n x^n,$$

and $a_i$, $i = 0 \ldots, n$ are real. The classical Hamilton's equations are:

$$\dot{x} = \frac{\partial H}{\partial p} = \frac{p}{m},$$

$$\dot{p} = -\frac{\partial H}{\partial p} = -\frac{dV}{dx}(x).$$

Recall the one dimensional momentum and position operators:

$$P = \frac{\hbar}{i}\frac{d}{dx}, \qquad X = x, \quad \text{(multiplication by } x\text{)}.$$

(b) Show that:

$$\frac{d}{dt}\langle X\rangle = \frac{\langle P\rangle}{m},$$

$$\frac{d}{dt}\langle P\rangle = -\left\langle \frac{dV}{dx}(X)\right\rangle.$$

These equations are known as Ehrenfest's equations.

(c) Under what conditions are Ehrenfest's equations of motion for $\langle X\rangle$ and $\langle P\rangle$ equivalent to the classical equations of motion for $x$ and $p$?

14. Let $|\psi_1\rangle$ and $|\psi_2\rangle$ denote orthonormal vectors in a state space. Let $A$ denote a self-adjoint linear operator on the state space and let $\alpha_n$ be a nondegenerate eigenvalue of $A$ with corresponding orthonormal eigenvector $|\phi_n\rangle$. Consider also the state:

$$|\psi\rangle = 3|\psi_1\rangle - 4i|\psi_2\rangle.$$

(a) What is the interpretation of the quantities $|\langle\psi_1|\phi_n\rangle|^2$ and $|\langle\psi_2|\phi_n\rangle|^2$?

(b) Normalize $|\psi\rangle$.

(c) Compute the probability of measuring $\alpha_n$ when $A$ is in the state $|\psi\rangle$.

(d) What is the state of the system after $\alpha_n$ is measured in the state $|\psi\rangle$?

15. Consider a particle of mass $m$ moving in the square well, i.e., in the interval $[0, a]$, where $V(x) = 0$ in this interval, and $V(x) = \infty$ for $x > a$ and $x < 0$.

The energy levels are given by:

$$E = E_n = \frac{n^2\pi^2\hbar^2}{2ma^2}, \quad n = 1, 2, \ldots,$$

and the corresponding eigenfunctions:

$$\psi_n(x) = \sqrt{\frac{2}{a}}\sin\left(\frac{n\pi x}{a}\right),$$

from which it follows that:

$$\psi_n(x, t) = \sqrt{\frac{2}{a}}\sin\left(\frac{n\pi x}{a}\right)e^{-\frac{in^2\pi^2\hbar t}{2ma^2}}.$$

(a) Compute $E_{\psi_n}(X)$, where $E_{\psi_n}(X)$ denotes the expectation value of $X$ in the state $\psi_n$.

(b) Compute $E_{\psi_n}(X^2)$.

(c) Compute $E_{\psi_n}(P)$.

(d) Compute $E_{\psi_n}(P^2)$.

(e) State the uncertainty relation and determine the state $\psi_n$ for which the uncertainty is a minimum.

16. Consider the quantum mechanical harmonic oscillator in one dimension described by the Hamiltonian:

$$H = \frac{1}{2m}P^2 + \frac{1}{2}m\omega^2 X^2,$$

where $P$ and $X$ are the momentum and positions operators, respectively.

Recall the associated raising and lowering operators:

$$a^\dagger = P + im\omega X,$$

$$a = P - im\omega X,$$

and the number operator

$$N = \frac{1}{2m\hbar\omega}a^\dagger a.$$

Let $|n\rangle$ denote the normalized eigenstates of $H$, $n = 0, 1, 2, \ldots$.
You may use the relations:

$$a^\dagger|n\rangle = \sqrt{2m\hbar\omega(n+1)}\,|n+1\rangle,$$

$$a|n\rangle = \sqrt{2m\hbar\omega n}\,|n-1\rangle.$$

Using $a^\dagger$ and $a$ show the following:

(a) Calculate $\langle n|X|n\rangle$.

(b) Calculate $\langle n|P|n\rangle$.

(c) Calculate $\langle n|X^2|n\rangle$.

(d) Calculate $\langle n|P^2|n\rangle$.

(e) Calculate $\Delta X \Delta P$ in the state $|n\rangle$, and determine the state for which it is the smallest.

17. Consider the quantum mechanical harmonic oscillator in one dimension described by the Hamiltonian:

$$H = \frac{1}{2m}P^2 + \frac{1}{2}m\omega^2 X^2,$$

where $P$ and $X$ are the momentum and positions operators, respectively.

Recall the associated raising and lowering operators:

$$a^\dagger = P + im\omega X,$$

$$a = P - im\omega X,$$

and the number operator

$$N = \frac{1}{2m\hbar\omega} a^\dagger a.$$

We define

$$\hat{a}(t) \equiv e^{\frac{i}{\hbar}Ht}\, a\, e^{-\frac{i}{\hbar}Ht}$$

(a) Show that

$$\hat{a}(t) \equiv e^{\frac{i}{\hbar}Ht}\, a\, e^{-\frac{i}{\hbar}Ht} = e^{i\omega Nt}\, a\, e^{-i\omega Nt}.$$

Justify all steps in your calculation.

(b) Show that

$$\frac{d}{dt}\hat{a}(t) = -i\omega\hat{a}(t), \qquad (\star)$$

(c) Solve the initial value problem $(\star)$ and show

$$\hat{a}(t) = e^{-i\omega t}\, a.$$

(d) Compute $\hat{a}^\dagger(t)$.

(e) Use $\hat{a}(t)$ and $\hat{a}^\dagger(t)$ to determine the time dependent position operator, $\hat{X}(t)$.

18. Consider the free particle on $\mathbb{R}$ of constant mass $m$. The Hamiltonian is given by:

$$H = \frac{P^2}{m},$$

where $P$ is the momentum operator.

(a) Write down Ehrenfest's equations for the one dimensional free particle, and solve them.

(b) Show that $\langle P^2 \rangle$ is constant.

(c) Show that

$$\frac{d}{dt}\langle (XP + PX) \rangle = \frac{2}{m}\langle P^2 \rangle.$$

(d) Show that

$$\frac{d}{dt}\langle X^2 \rangle = \frac{1}{m}\langle (XP + PX) \rangle.$$

(e) Discuss the time dependence of $(\Delta P)^2$.

# Chapter 4

# Quantum Mechanics of Angular Momentum

## 4.1  Angular Momentum: Classical to Quantum

Why is angular momentum interesting? We are not going to spend a great deal of time answering this question, but you should think about it and come up with your own answers. However, you could ask the same question of "energy" and "linear momentum". One answer might be that they have proven to be very useful quantities for describing the behaviour of mechanical systems (at this point we have not distinguished classical and quantum mechanical systems – you should think about that). This is true. However, you might think of some simple mechanical systems (e.g. a point mass vibrating at the end of a spring, a point mass on the end of a string that you are swinging in a circle) and think about what quantities would best describe the motion of that system. Once you have answered this question to your satisfaction, then you could go on to a "deeper" question (that's what scientists do), and you could ask "where do energy, linear momentum, and angular momentum come from?" It's not entirely clear what would comprise a satisfactory answer to this question. However, one important, and deep, answer to this question lies in understanding the notion of *symmetry* and it's role in mechanics, both classical and quantum. There is a remarkable theorem proven by the mathematician Emmy Noether called (not surprisingly) "Noether's theorem" that relates specific symmetries to "conserved quantities" (such as energy, linear momentum, and angular momentum).[1] This will not be discussed in this book, but if you go further in your studies of classical and quantum mechanics, you will undoubtedly encounter this topic. For now, we will begin with the "traditional" way of introducing angular momentum in quantum mechanics. The development of angular momentum in this chapter is very much the same as Littlejohn's "Notes 13".

We begin with the expression for the *classical* angular momentum of a single particle about a point (which we will refer to as "the origin", if required):

$$\mathbf{L} = \mathbf{x} \times \mathbf{p} \tag{4.1}$$

where $\mathbf{x} \equiv (x_1, x_2, x_3)$ is the position of the particle with respect to the origin and $\mathbf{p} \equiv (p_1, p_2, p_3)$ is the (linear) momentum of the particle. The components of the angular momentum vector are given by:

$$L_1 = x_2 p_3 - x_3 p_2,$$

$$L_2 = x_3 p_1 - x_1 p_3,$$

$$L_3 = x_1 p_2 - x_2 p_1, \tag{4.2}$$

Now we will introduce some notation that will enable us to more concisely write the expression for angular momentum. It will also enable us to much more simply manipulate various quantities associated with angular momentum (both classical and quantum).

The Levi-Civita symbol (sometimes also referred to as the "permutation symbol", "antisymmetric symbol", or "alternating symbol") is defined as:

$$\epsilon_{ijk} = \begin{cases} 1 & \text{if } (i,j,k) \text{ is } (1,2,3), (3,1,2), \text{ or } (2,3,1), \\ -1 & \text{if } (i,j,k) \text{ is } (1,3,2), (3,2,1), \text{ or } (2,1,3), \\ 0 & \text{if } i=j \text{ or } j=k \text{ or } k=i. \end{cases} \tag{4.3}$$

Stated more concisely, $\epsilon_{ijk}$ is 1 if $(i,j,k)$ is an even permutation of $(1,2,3)$, $-1$ if it is an odd permutation, and 0 if any index is repeated.

Using the Levi-Civita symbol we have:

$$L_i = \sum_{j=1}^{3} \sum_{k=1}^{3} \epsilon_{ijk} x_j p_k, \qquad i = 1, 2, 3. \tag{4.4}$$

---

[1] Emmy Noether was one of the greatest mathematicians of the twentieth century. The breadth of her contributions to mathematics is truly remarkable. Her Wikipedia entry gives a nice overview of her life, https://en.wikipedia.org/wiki/Emmy_Noether. Recently, a book has been published solely devoted to Noether's theorem.

Dwight E Neuenschwander. *Emmy Noether's wonderful theorem.* JHU Press, 2017.

Of course, the first time you see this, you really should write it out completely so that you are sure that the compact notation really does give (4.2). Doing this gives:

$$L_i = \sum_{j=1}^{3}\sum_{k=1}^{3} \epsilon_{ijk} x_j p_k$$

$$= \sum_{j=1}^{3}\left( \epsilon_{ij1} x_j p_1 + \epsilon_{ij2} x_j p_2 + \epsilon_{ij3} x_j p_3 \right)$$

$$= \epsilon_{i11} x_1 p_1 + \epsilon_{i21} x_2 p_1 + \epsilon_{i31} x_3 p_1$$

$$+ \epsilon_{i12} x_1 p_2 + \epsilon_{i22} x_2 p_2 + \epsilon_{i32} x_3 p_2$$

$$+ \epsilon_{i13} x_1 p_3 + \epsilon_{i23} x_2 p_3 + \epsilon_{i33} x_3 p_3$$

$$= 0 + \epsilon_{i21} x_2 p_1 + \epsilon_{i31} x_3 p_1$$

$$+ \epsilon_{i12} x_1 p_2 + 0 + \epsilon_{i32} x_3 p_2$$

$$+ \epsilon_{i13} x_1 p_3 + \epsilon_{i23} x_2 p_3 + 0. \tag{4.5}$$

From which it follows immediately that

$$L_1 = \epsilon_{132} x_3 p_2 + \epsilon_{123} x_2 p_3 = x_2 p_3 - x_3 p_2, \tag{4.6}$$

$$L_2 = \epsilon_{231} x_3 p_1 + \epsilon_{213} x_1 p_3 = x_3 p_1 - x_1 p_3, \tag{4.7}$$

$$L_3 = \epsilon_{321} x_2 p_1 + \epsilon_{312} x_1 p_2 = x_1 p_2 - x_2 p_1. \tag{4.8}$$

Now that we are sure that (4.4) "encodes" (4.2) we can simplify the notion even further. Instead of writing the "double sum" in (4.2) we write:

$$L_i = \sum_{jk} \epsilon_{ijk} x_j p_k, \quad i = 1, 2, 3. \tag{4.9}$$

We can simplify the expression further by writing (4.9) as:

$$L_i = \epsilon_{ijk} x_j p_k, \quad i = 1, 2, 3, \tag{4.10}$$

where it is understood that repeated indices are summed from 1 to 3. This "summation over repeated indices" is referred to as the "Einstein summation convention". To summarize, the compact notation of (4.10) means the same as (4.4). We will use the Einstein summation convention throughout these notes.

Now passing to the quantum mechanical version of (orbital) angular momentum, we apply the quantization rules of associating to the position variables $x_1$, $x_2$, $x_3$ the self-adjoint position operators $X_1$, $X_2$, $X_3$, respectively, and to the momentum variables $p_1$, $p_2$, $p_3$ the self-adjoint momentum operators $P_1$, $P_2$, $P_3$, respectively and form the quantities leading to the following definition.

**Definition 27 (Quantum Mechanical (Orbital) Angular Momentum).**

$$L_1 = X_2 P_3 - X_3 P_2, \tag{4.11}$$

$$L_2 = X_3 P_1 - X_1 P_3, \tag{4.12}$$

$$L_3 = X_1 P_2 - X_2 P_1, \tag{4.13}$$

which we denote more concisely using the Levi-Civita symbol as follows:

$$L_i = \epsilon_{ijk} X_j P_k, \quad i = 1, 2, 3. \tag{4.14}$$

The operators $L_i$, $i = 1, 2, 3$ defined in this way are self-adjoint, as we now show.

**Proposition 1.** The operators $L_1$, $L_2$, $L_3$ are self-adjoint operators.

**Proof.** Initially, one might think that this statement is "obvious" since the $L_i$ are the sum of the product of two self-adjoint operators. However, it is not necessarily true that the product of two self-adjoint operators is self-adjoint, and in the course of showing that the $L_i$ are self-adjoint, we will see the crucial issue.

We carry out the proof for $L_1 = X_2 P_3 - X_3 P_2$. The proof for $L_2$ and $L_3$ is analogous. We have

$$L_1^\dagger = (X_2 P_3 - X_3 P_2)^\dagger = (X_2 P_3)^\dagger - (X_3 P_2)^\dagger = P_3^\dagger X_2^\dagger - P_2^\dagger X_3^\dagger$$

$$= P_3 X_2 - P_2 X_3. \tag{4.15}$$

Now, if we could "reverse the order" of $P_3 X_2$ and $P_2 X_3$ we would be finished. In other words, is it true that $P_3 X_2 = X_2 P_3$ and $P_2 X_3 = X_3 P_2$? In other words, do $X_2$ and $P_3$ commute, and do $X_3$ and $P_2$ commute? The answer is "yes". Recall the commutation relations:

$$[X_j, P_k] = i\hbar \delta_{j,k}, \quad [X_j, X_k] = [P_j, P_k] = 0, \quad j, k = 1, 2, 3. \tag{4.16}$$

Hence, we have

$$L_1^\dagger = (X_2 P_3 - X_3 P_2)^\dagger = P_3 X_2 - P_2 X_3 = X_2 P_3 - X_3 P_2 = L_1. \tag{4.17}$$

$$\square$$

The next result concerns the commutation relations of (orbital) angular momentum and will be fundamental to the rest of the material on angular momentum.

## Proposition 2.

1. $[L_j, P_k] = i\hbar\epsilon_{jkl}P_l$
2. $[L_j, X_k] = i\hbar\epsilon_{jkl}X_l$
3. $[L_j, L_k] = i\hbar\epsilon_{jkl}L_l$

Before proving this result we need a preliminary lemma on commutators in general.

Lemma 1. For any operators $A$, $B$, $C$,

$$i)\; [AB, C] = A[B, C] + [A, C]B \qquad ii)\; [C, AB] = A[C, B] + [C, A]B$$

**Proof.** The proof uses the basic "trick" of adding and subtracting the same quantity to an expression, but combining the two quantities with different terms in the expression. For i) we have:

$$[AB, C] = ABC - CAB$$
$$= A\left(BC - CB + CB\right) - CAB$$
$$= A\left([B, C] + CB\right) - CAB$$
$$= A[B, C] + \left(AC - CA + CA\right)B - CAB$$
$$= A[B, C] + \left([A, C] + CA\right)B - CAB$$
$$= A[B, C] + [A, C]B$$

The proof of ii) follows immediately if we note that $[AB, C] = -[C, AB]$.
$$\square$$

We now return to the proof of Proposition 2.

**Proof.** We begin with 1.

$$[L_j, P_k] = [\epsilon_{jmn}\, X_m P_n, P_k]$$
$$= \epsilon_{jmn}\, [X_m P_n, P_k]$$
$$= \epsilon_{jmn}\, \left(X_m[P_n, P_k] + [X_m, P_k]P_n\right) \quad \text{using Lemma 1 i)}$$
$$= \epsilon_{jmn}\, \left(0 + i\hbar\delta_{mk}P_n\right) \quad \text{using (4.16)}$$
$$= i\hbar\epsilon_{jkn}\, P_n$$

The proof of 2 proceeds along the same lines.

$$[L_j, X_k] = [\epsilon_{jmn} X_m P_n, X_k]$$

$$= \epsilon_{jmn} [X_m P_n, X_k]$$

$$= \epsilon_{jmn} \left( X_m[P_n, X_k] + [X_m, X_k]P_n \right) \quad \text{using Lemma 1 i)}$$

$$= \epsilon_{jmn} \left( -i\hbar\delta_{kn}X_m + 0 \right) \quad \text{using (4.16)}$$

$$= -i\hbar\epsilon_{jmk} X_m,$$

$$= i\hbar\epsilon_{jkm} X_m$$

We prove 3 for $L_1$ and $L_2$. The proof for the other components is easily obtained by permuting the indices appropriately.

$$[L_1, L_2] = [L_1, X_3 P_1 - X_1 P_3]$$

$$= [L_1, X_3 P_1] - [L_1, X_1 P_3]$$

$$= X_3[L_1, P_1] + [L_1, X_3]P_1 - X_1[L_1, P_3] - [L_1, X_1]P_3$$

$$\text{using Lemma 1 ii)}$$

$$= 0 + i\hbar\epsilon_{13l}X_l P_1 - X_1 i\hbar\epsilon_{13l}P_l + 0$$

$$\text{using parts 1 and 2 of this Proposition}$$

$$= i\hbar\epsilon_{132} \left( X_2 P_1 - X_1 P_2 \right)$$

$$= i\hbar(X_1 P_2 - X_2 P_1) = i\hbar L_3 \qquad \qquad \square$$

The commutation relations in statement 3 of Proposition 2 are really the "heart" of the quantum theory of angular momentum. You will see that they form the basis of essentially all of the results that we develop related to angular momentum. In more advanced treatments of this topic it can be shown that operators satisfying these commutation relations have their origins in the properties of rotations in three dimensional space. Therefore it is natural to define any three self-adjoint operators that satisfy such commutation related as quantum angular momentum operators.

**Definition 28 (General Definition of Quantum Angular Momentum Operators).** Any three self-adjoint operators $J_1$, $J_2$, $J_3$ that satisfy:

$$[J_j, J_k] = i\hbar\epsilon_{jkl} J_l, \tag{4.18}$$

are called (quantum) angular momentum operators.

An obvious question is "why do we do this"? The quantization of classical orbital angular momentum was fairly straightforward. Why does this need to be "generalized" in any way? It turns out that there are other "types of angular momentum", in addition to the quantization of classical orbital angular momentum. "Spin" is one type of "other" angular momentum. We will study spin explicitly at the end of our study of angular momentum. However, for now we merely state that spin is a property of particles (such as electrons, protons, neutrons, photons, etc) that has no classical analog. It is however described by three self-adjoint operators satisfying the same commutation relations that we derived for orbital angular momentum. This is a good place to say something about terminology.

> **Terminology.** Eigenvalues and eigenvectors are common terms associated with linear operators. However, in quantum mechanics some synonyms are associated with the term "eigenvector" that are specific to the particular physics and/or notation context. For example, the word "eigenstate" or "eigenket" (reflecting the influence of the Dirac notation) is often used synonomously for the term "eigenvector".

Our next goal is to construct a basis of angular momentum eigenvectors (what does this mean?). Initially, this would seem to pose a problem since we cannot construct a basis consisting of simultaneous eigenvectors of $J_1$, $J_2$, and $J_3$ since these three operators do not commute with each other (we will comment on this shortly). However, we can motivate this using an analogy with classical orbital angular momentum. Classical orbital angular momentum is a vector in three dimensions, and therefore it can be described by a magnitude and a direction. Quantum mechanically we can think of the operator defined by the square of the vector of three operators defining the quantum mechanical angular momentum, i.e. the square of

$$\mathbf{J} \equiv (J_1, J_2, J_3),$$

as defining the square of the magnitude of total angular momentum:

$$\mathbf{J} \cdot \mathbf{J} \equiv \mathbf{J}^2 = J_1^2 + J_2^2 + J_3^2. \tag{4.19}$$

It is easy to see that $\mathbf{J}^2$ is self-adjoint since it is the sum of the squares of three self-adjoint operators (we don't have the problem that we had earlier since $J_i$ commutes with $J_i$, $i = 1, 2, 3$). Moreover, we have the following result.

**Proposition 3.**

$$[\mathbf{J}^2, J_i] = 0, \quad i = 1, 2, 3.$$

## Proof.

$$[\mathbf{J}^2, J_i] = [J_j J_j, J_i],$$

$$= J_j[J_j, J_i] + [J_j, J_i]J_j, \quad \text{using Lemma 1 i)}$$

$$= i\hbar J_j \epsilon_{jik} J_k + i\hbar \epsilon_{jik} J_k J_j, \quad \text{using (4.18)}$$

$$= i\hbar \epsilon_{jik} (J_j J_k + J_k J_j)$$

$$= 0,$$

since $\epsilon_{jik}$ is antisymmetric in $j, k$ and the term in parentheses is symmetric in $j, k$. We show this last step explicitly. We have

$$\epsilon_{jik} J_k J_j = \epsilon_{kij} J_j J_k,$$

where we have just interchanged the summation indices $k$ and $j$.

$$= -\epsilon_{jik} J_j J_k,$$

where in the last step we have used the fact that antisymmetry implies we can interchange the order of two indices, but this requires us to multiply the resulting term by a minus sign. Adding the left and right hand sides gives:

$$\epsilon_{jik} J_j J_k + \epsilon_{jik} J_k J_j = \epsilon_{jik}(J_j J_k + J_k J_j) = 0. \qquad \square$$

This result implies that we can find simultaneous eigenvectors of $\mathbf{J}$ and one component of angular momentum, which we shall take as $J_3$ (mainly for traditional reasons). Now there is a *lot* in this result, and in the statements leading up to it, that we now want to discuss.

Recall that two of the nicest and most useful properties of self-adjoint operators are that their eigenvalues are real, and eigenvectors corresponding to distinct eigenvalues are orthogonal (and we will always be normalizing our eigenvectors so that they have unit length – if this is not familiar, you should go back and review this from earlier in the book). It could happen that two different eigenvectors have the same eigenvalue. This situation is referred to as a *degeneracy*. We will not deal with degeneracies explicitly at the moment, but it is important to realize that they can arise and whether or not our algebraic manipulations allow for their possibility. Even when degeneracies exist, it is still possible to find a "complete set" of orthonormal eigenvectors. You just have to work a bit harder.

Now let's consider two self-adjoint operators, and let's rule out the possibility of degeneracies for the moment (it makes getting the main idea across easier – we can come back and consider the case of degeneracies afterwards). We know that each self-adjoint operator, individually, possesses a *complete set of orthonormal eigenvectors* (make sure you know what the word "complete" means in this statement). However, if the two operators commute (and there are no degeneracies, for the moment) then it is possible to find a complete set of orthonormal vectors that are eigenvectors for each self-adjoint operator simultaneously. Commutation of the operators is essential here. So, getting back to angular momentum, we know that there exists a complete set of orthonormal vectors that are eigenvectors for both $\mathbf{J}^2$ and $J_3$. We just need to be able to compute them, which is the next topic. The same result can be shown to hold when there are degeneracies, but for now we will postpone dealing with that situation. The book by Cohen-Tannoudji et al.[2] has a very nice discussion of these ideas in the chapter on "Mathematical Tools of Quantum Mechanics". The Littlejohn notes also have a good treatment of this topic. (Recall that we proved this result earlier in the book, and even dealt with the case of degeneracies.)

Before going on to compute eigenvalues and eigenvectors associated with angular momentum, it is worth making some final remarks. Recall that in quantum mechanics self-adjoint operators are referred to as "observables". They are the mathematical manifestations of quantities that we can "observe", i.e. measure. The latter part of this book will be very much concerned with this topic. However, recall that two observables can be measured simultaneously if and only if the corresponding operators commute. So we have shown that $\mathbf{J}^2$ commutes with $J_1$, $J_2$ and $J_3$. Does this mean that $\mathbf{J}^2$, $J_1$, $J_2$, and $J_3$ can all be measured simultaneously? But $J_1$, $J_2$, and $J_3$ do not commute. What is the situation here?

## 4.2 Eigenvalues and Eigenvectors of Angular Momentum

Now we are going to do three things.

- Compute eigenvalues of $\mathbf{J}^2$ and $J_3$.
- Compute eigenvectors that are simultaneously eigenvectors of $\mathbf{J}^2$ and $J_3$.
- Use the eigenvectors to compute matrix representations of the components of $\mathbf{J}$ for "certain physical situations" (which will be more fully described when we get to it).

---

[2]C. Cohen-Tannoudji, B. Diu, and F. Laloe. *Quantum Mechanics.* Wiley-VCH, 1992.

The material below follows closely the "Littlejohn Notes" ("Notes 13"). We denote the eigenvalues of $\mathbf{J}^2$ by $a\hbar^2$ and the eigenvalues of $J_3$ by $m\hbar$. The factors of $\hbar$ may seem a bit mysterious. However, recall that $\hbar$ has the units of angular momentum, i.e. Joule $\cdot$ second or kilogram $\cdot$ meter$^2$ /second. Therefore the factors of $\hbar$ serve to make the eigenvalues $a$ and $m$ dimensionless. We will label the eigenvectors by the eigenvalues $a$ and $m$ as $|am\rangle$, and therefore we have:

$$\mathbf{J}^2|am\rangle = a\hbar^2|am\rangle,$$

$$J_3|am\rangle = m\hbar|am\rangle. \tag{4.20}$$

We are assuming nondegeneracy, so that $a$ and $m$ are unique labels for the simultaneous eigenvectors of $\mathbf{J}$ and $J_3$, up to normalization and the choice of a phase. However, we will always assume that the eigenvectors are normalized, i.e.,

$$\langle am|am\rangle = 1. \tag{4.21}$$

Later we will consider the consequences of allowing for degeneracies.

The following operators will play an important role in our construction of eigenvalues and eigenvectors.

**Definition 29 (Ladder Operators for Quantum Angular Momentum).** The "ladder operators" $J_+$ and $J_-$ are defined by:

$$J_+ = J_1 + iJ_2, \tag{4.22}$$

$$J_- = J_1 - iJ_2 \tag{4.23}$$

It should be clear that these operators are self-adjoint conjugates in the sense that:

$$(J_\pm)^\dagger = J_\mp. \tag{4.24}$$

The following result gives some fundamental commutation relations among $\mathbf{J}^2$, $J_3$ and the ladder operators that will play important roles in our understanding of the "eigenstructure" of $\mathbf{J}^2$ and $J_3$.

**Proposition 4.**

1. $[J_3, J_\pm] = \pm\hbar J_\pm,$
2. $[J^2, J_\pm] = 0,$
3. $[J_+, J_-] = 2\hbar J_3,$
4. $J_\pm J_\mp = J^2 - J_3^2 \pm \hbar J_3.$

**Proof.** The proofs of these results use the definition of the ladder operators (Definition 29), the angular momentum commutation relations given in (4.18), and Proposition 3. The necessary calculations proceed as follows.

1.
$$[J_3, J_\pm] = [J_3, J_1 \pm iJ_2]$$
$$= [J_3, J_1] \pm i[J_3, J_2]$$
$$= i\hbar J_2 \pm i(-i\hbar)J_1$$
$$= i\hbar J_2 \pm \hbar J_1$$
$$= \pm\hbar(J_1 \pm iJ_2)$$
$$= \pm\hbar J_\pm$$

2.
$$[J^2, J_\pm] = [J^2, J_1 \pm iJ_2]$$
$$= [J^2, J_1] \pm i[J^2, J_2]$$
$$= 0 \pm i\,0 = 0$$

3.
$$[J_+, J_-] = [J_1 + iJ_2, J_1 - iJ_2]$$
$$= [J_1 + iJ_2, J_1] + [J_1 + iJ_2, -iJ_2]$$
$$= [J_1, J_1] + [iJ_2, J_1] + [J_1, -iJ_2] + [iJ_2, -iJ_2]$$
$$= 0 + i(-i\hbar)J_3 - i(i\hbar)J_3 + 0$$
$$= 2\hbar J_3$$

4.
$$J_+ J_- = (J_1 + iJ_2)(J_1 - iJ_2)$$
$$= J_1^2 + J_2^2 - i(J_1 J_2 - J_2 J_1)$$
$$= J^2 - J_3^2 - i[J_1, J_2]$$
$$= J^2 - J_3^2 + \hbar J_3 \tag{4.25}$$
$$J_- J_+ = (J_1 - iJ_2)(J_1 + iJ_2)$$
$$= J_1^2 + J_2^2 + i(J_1 J_2 - J_2 J_1)$$
$$= J^2 - J_3^2 + i[J_1, J_2]$$
$$= J^2 - J_3^2 - \hbar J_3 \tag{4.26}$$

$\square$

We will also need the results given in the following Corollary.

**Corollary 1.**
$$\langle am|J_- J_+|am\rangle = \hbar^2\left(a - m(m+1)\right) \geq 0 \tag{4.27}$$
$$\langle am|J_+ J_-|am\rangle = \hbar^2\left(a - m(m-1)\right) \geq 0 \tag{4.28}$$

**Proof.** We first prove (4.27).

Using (4.26), we have

$$\langle am|J_-J_+|am\rangle = \langle am|\left(J^2 - J_3^2 - \hbar J_3\right)|am\rangle$$

$$= \langle am|J^2|am\rangle - \langle am|J_3^2|am\rangle - \hbar\langle am|J_3|am\rangle$$

$$= \hbar^2 a\langle am|am\rangle - \hbar^2 m^2\langle am|am\rangle - \hbar^2 m\langle am|am\rangle$$

$$= \hbar^2\left(a - m(m+1)\right)\langle am|am\rangle$$

$$= \hbar^2\left(a - m(m+1)\right)$$

It remains to argue that this expression is nonnegative. Using (4.24), we have

$$(J_+|am\rangle)^\dagger = \langle am|J_+^\dagger = \langle am|J_-,$$

and therefore $\langle am|J_-J_+|am\rangle$ is the square of the norm of $J_+|am\rangle$, which is clearly nonnegative.

The proof of (4.28) is completely analogous to the proof of (4.27). $\quad\square$

Now we have the tools to prove our main result.

First, note that from (4.27) and (4.28) that we have:

$$a \geq \max\left[m(m+1), m(m-1)\right]. \tag{4.29}$$

The functions $m(m-1)$ and $m(m+1)$ are plotted in Fig. 4.1(a), and the maximum of these functions is plotted in Fig. 4.1(b).

It is not hard to verify that the maximum function is symmetric about $m = 0$ and $\geq 0$ everywhere. We choose a value of $a \geq 0$ and draw a horizontal line in Fig. 4.1(b) corresponding to $a = \max\left[m(m+1), m(m-1)\right]$. This shows that for any $a \geq 0$ there is a maximum and a minimum value of $m$ (which, by symmetry, are of equal magnitude) for which (4.29) is satisfied for all values in between the maximum and minimum. For $a \geq 0$ we denote the maximum value of $m$ by $j$ and the minimum value of $m$ by $-j$. Therefore we have:

$$-j \leq m \leq j, \tag{4.30}$$

Clearly, $j$ is a function of $a$, and $j \geq 0$ since $a \geq 0$. From Fig. 4.1(b) we see that:

$$a = j(j+1). \tag{4.31}$$

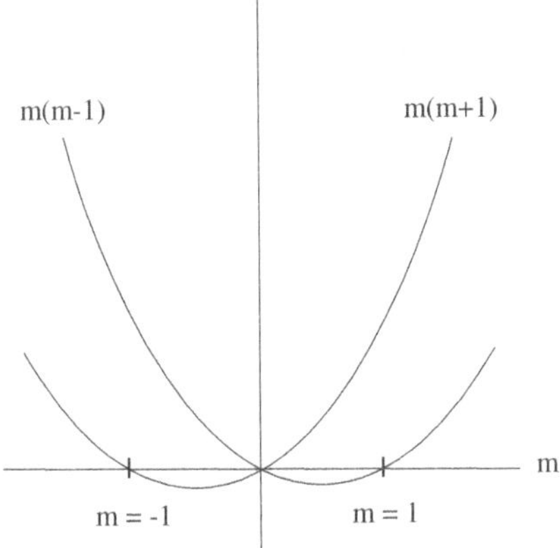



(a) Graphs of the functions $m\,(m-1)$ and $m\,(m+1)$.
(Adapted from Littlejohn "Notes 13".)

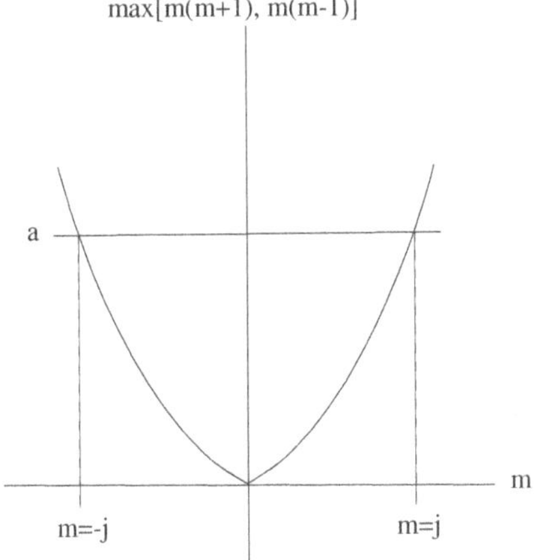



(b) Graph of the function max $(m(m+1), m(m-1))$ with maximum
and minimum values of $m$ shown for a given value of $a$.
(Adapted from Littlejohn "Notes 13".)

We will find it more convenient to parametrize the eigenvectors of $\mathbf{J}^2$ by $j$, rather than $a$. Therefore we will write $j(j+1)$ for the eigenvalue of $\mathbf{J}^2$, rather than $a$, and we will denote the simultaneous eigenvectors of $\mathbf{J}^2$ and $J_3$ by $|jm\rangle$, rather than $|am\rangle$. Then (4.20) is rewritten as follows:

$$\mathbf{J}^2|jm\rangle = j(j+1)\hbar^2|jm\rangle,$$

$$J_3|jm\rangle = m\hbar|jm\rangle. \tag{4.32}$$

Using this new notation, we also rewrite (4.27) and (4.28) as (where we have also factored the right hand side of each expression):

$$\langle jm|J_-J_+|jm\rangle = \hbar^2\left(j(j+1) - m(m+1)\right)$$

$$= \hbar^2(j-m)(j+m+1) \geq 0, \tag{4.33}$$

$$\langle jm|J_+J_-|jm\rangle = \hbar^2\left(j(j+1) - m(m-1)\right)$$

$$= \hbar^2(j+m)(j-m+1) \geq 0, \tag{4.34}$$

Now we want to consider the conditions under which the inequalities (4.33) and (4.34) become equalities, i.e. under what conditions do the matrix elements on the left hand side of the equals sign become zero? First we consider (4.33). In this case we have $J_+|jm\rangle = 0$ if and only if

$$j - m = 0 \quad \text{or} \quad j + m + 1 = 0, \tag{4.35}$$

or, equivalently

$$m = j \quad \text{or} \quad m = -j - 1, \tag{4.36}$$

However, from (4.30), it follows that $m = -j - 1$ cannot occur, and therefore,

$$J_+|jm\rangle = 0 \quad \text{if and only if} \quad m = j. \tag{4.37}$$

A similar analysis of (4.34) shows that:

$$J_-|jm\rangle = 0 \quad \text{if and only if} \quad m = -j. \tag{4.38}$$

We assume that $|jm\rangle$ is a normalized eigenvector of $\mathbf{J}^2$ and $J_3$ with eigenvalues $j(j+1)\hbar^2$ and $m\hbar$. Using Proposition 4, part 1, we have:

$$J_3\left(J_\pm|jm\rangle\right) = \left(J_\pm J_3 \pm \hbar J_\pm\right)|jm\rangle = m\hbar J_\pm|jm\rangle \pm \hbar J_\pm|jm\rangle$$

$$= (m \pm 1)\hbar\left(J_\pm|jm\rangle\right), \tag{4.39}$$

and, using Proposition 4, part 2 we have

$$\mathbf{J}^2\left(J_\pm|jm\rangle\right) = J_\pm\mathbf{J}^2|jm\rangle = j(j+1)\hbar^2\left(J_\pm|jm\rangle\right). \tag{4.40}$$

From these two calculations we make the following conclusions.

- If $J_+|jm\rangle$ does not vanish, then it is an eigenvector of $\mathbf{J}^2$ and $J_3$ with eigenvalues $j(j+1)\hbar^2$ and $(m+1)\hbar$, respectively. In other words, $J_+$ does not change $j$, but it increases $m$ by 1.
- If $J_-|jm\rangle$ does not vanish, then it is an eigenvector of $\mathbf{J}^2$ and $J_3$ with eigenvalues $j(j+1)\hbar^2$ and $(m-1)\hbar$, respectively. In other words, $J_-$ does not change $j$, but it decreases $m$ by 1.

Therefore, repeated application of $J_+$ on $|jm\rangle$ generates the sequence of eigenvectors:

$$|jm\rangle, \ |j, m+1\rangle. \ \ldots, \ |j, m+n_1\rangle,$$

which must terminate for some integer $n_1 \geq 0$, or else the bounds (4.30) would be violated. Hence we have:

$$j = m + n_1, \quad \text{or} \quad m = j - n_1. \tag{4.41}$$

Similarly, repeated application of $J_-$ on $|jm\rangle$ generates the sequence of eigenvectors:

$$|jm\rangle, \ |j, m-1\rangle. \ \ldots, \ |j, m-n_2\rangle,$$

which must terminate for some integer $n_2 \geq 0$, or else the bounds (4.30) would be violated. Hence we have:

$$-j = m - n_2, \quad \text{or} \quad m = -j + n_2. \tag{4.42}$$

Taken together, (4.41) and (4.42) imply that:

$$2j = n_1 + n_2 \geq 0, \tag{4.43}$$

This implies that the only allowed values of $j$ are:

$$j \in \left\{ 0, \frac{1}{2}, 1, \frac{3}{2}, \ldots \right\} \tag{4.44}$$

> **Key point:** It is important to realize that (4.44) only tells us that these are the values of $j$ allowed by the angular momentum commutation relations. It does not tell us which particular values occur in a specific application. This highlights the strength of considering "angular momentum in general, as defined by the commutation relations, and then worrying about the particular type of angular momentum when considering specific applications.
>
> For example, when the angular momentum under consideration corresponds to spin (something we have not considered yet) a "spin $\frac{1}{2}$" particle only has $j = \frac{1}{2}$. For a particle moving in a central force potential orbital angular momentum is the relevant type of angular momentum, and it can only possess eigenvalues having integer values of $j$ (which we will discuss shortly).

We summarize the main result of this section as follows.

**Theorem 10.** *The operator $\mathbf{J}^2$ has eigenvalues $j(j+1)\hbar^2$, with $j \in \{0, \frac{1}{2}, 1, \frac{3}{2}, \ldots\}$. For each $j$ we have $2j+1$ eigenvectors, $|jm\rangle$, $m = -j, -j+1, \ldots, j-1, j$ which are eigenvectors of $J_3$ with respective eigenvalue $m\hbar$, i.e.*

$$\mathbf{J}^2|jm\rangle = j(j+1)\hbar^2|jm\rangle, \qquad J_3|jm\rangle = m\hbar|jm\rangle. \tag{4.45}$$

### *The Eigenvalues of $J_\pm$ and a Phase Convention*

We have shown that $J_+|jm\rangle$ and $J_-|jm\rangle$ are eigenvectors of $J_3$ with eigenvalues $(m+1)\hbar$ and $(m-1)\hbar$, respectively. Assuming nondegeneracy (why?), we can write

$$J_+|jm\rangle = c|j, m+1\rangle \tag{4.46}$$

$$J_-|jm\rangle = c'|j, m-1\rangle, \tag{4.47}$$

where $c$ and $c'$ are complex numbers that are arbitrary and therefore are chosen to be real and positive. The specific choices are a matter of convention, and are chosen as follows. Recalling (4.33) and (4.34), we have:

$$|c|^2\langle j, m+1|j, m+1\rangle = \langle jm|J_-J_+|jm\rangle$$

$$= |c|^2 = \hbar^2(j-m)(j+m+1)$$

$$|c'|^2\langle j, m-1|j, m-1\rangle = \langle jm|J_+J_-|jm\rangle$$

$$= |c'|^2 = \hbar^2(j+m)(j-m+1)$$

Hence, with the chosen phase convention, we can take:

$$c = \hbar\sqrt{(j-m)(j+m+1)}, \tag{4.48}$$

$$c' = \hbar\sqrt{(j+m)(j-m+1)}, \tag{4.49}$$

and therefore we have:

$$J_+|jm\rangle = \hbar\sqrt{(j-m)(j+m+1)}|j, m+1\rangle \tag{4.50}$$

$$J_-|jm\rangle = \hbar\sqrt{(j+m)(j-m+1)}|j, m-1\rangle \tag{4.51}$$

### Degeneracies and Multiplicities

In several places throughout our discussions we have explicitly assumed that there were "no degeneracies". Now we will consider this question more carefully (look in the section just above and convince yourself why the assumption of nondegeneracy was necessary).

First, we recall what it means for the simultaneous eigenvectors of $\mathbf{J}^2$ and $J_3$ to be degenerate, but now in a bit more detail. We denote the eigenspace of $\mathbf{J}^2$ and $J_3$ corresponding to eigenvalues $j(j+1)\hbar^2$ and $m\hbar$, respectively, by $\mathcal{E}_{jm}$ and we denote its dimension by:

$$\dim \mathcal{E}_{jm} = N_{jm}. \tag{4.52}$$

Then the simultaneous eigenvectors of $\mathbf{J}^2$ and $J_3$ are said to be degenerate if $N_{jm} > 1$. Stated another way, for a fixed $j$ and $m$, there is more than one simultaneous eigenvector of $\mathbf{J}^2$ and $J_3$ corresponding to that fixed $j$ and $m$. Here is the main general result on degeneracies for $\mathbf{J}^2$ and $J_3$.

*For fixed $j$ all the eigenspaces $\mathcal{E}_{jm}$ for $m = -j, \ldots, +j$ have the same dimension. We denote this dimension by $N_j$, which is referred to as the multiplicity of the $j$ value. The multiplicity can take on any value from 0 (which corresponds to that case of that particular $j$ value not occurring) to $\infty$.*

We give an outline of the proof that is almost exactly that given in the "Littlejohn notes", but we made a few comments considering where more details could be added.

The ladder operators will play an important role in this argument.

We start with the eigenspace $\mathcal{E}_{jj}$ which has dimension $N_{jj}$, and then we choose a set of $N_{jj}$ linearly independent vectors in this space (why can we make such a choice?). If we apply $J_-$ to these vectors, we obtain a set of $N_{jj}$ vectors that are eigenvectors of $\mathbf{J}^2$ and $J_3$ with eigenvalues $j(j+1)\hbar^2$ and $(m-1)\hbar$ (that is, with a lowered value of $m$). These vectors must lie in the eigenspace $\mathcal{E}_{j,j-1}$, and, as one can show, they are also linearly independent (you should show that this is true). Thus, $\dim \mathcal{E}_{j,j-1} = N_{j,j-1} \geq N_{jj}$ (why do we have $\geq$ here and not $=$?).

Now we go "back up" with the other ladder operator. Choose a set of $N_{j,j-1}$ linearly independent vectors in $\mathcal{E}_{j,j-1}$ and apply the raising operator $J_+$ to them. This creates a set of $N_{j,j-1}$ vectors that lie in the eigenspace $\mathcal{E}_{jj}$, which, as one can show, are also linearly independent (you should understand how this can be shown). Thus, $N_{jj} \geq N_{j,j-1}$. But this is consistent with $N_{j,j-1} \geq N_{jj}$ only if $N_{j,j-1} = N_{jj}$.

This argument can be repeated for all of the $m$ values, and in this way we see that all the eigenspaces $\mathcal{E}_{jm}$ for $m = -j, \ldots, j$ have the same dimension. We denote this dimension by $N_j$, which we call the multiplicity of the given $j$ value. The multiplicity can take on any value from 0 (in which case the $j$ value does not occur) to $\infty$.

## 4.3 Examples of Matrix Representations for Angular Momentum

Using eigenvectors we can construct matrix representations for systems with a fixed $j$.

First, we collect together some matrix elements that we have already computed (you need to go back in the notes and make sure of this) and that we will need.

$$\langle jm'|J_3|jm \rangle = m\hbar\delta_{m'm}, \tag{4.53}$$

$$\langle jm'|\mathbf{J}^2|jm \rangle = \hbar^2 j(j+1)\delta_{m'm}, \tag{4.54}$$

$$\langle jm'|J_+|jm \rangle = \hbar\sqrt{(j-m)(j+m+1)}\delta_{m',m+1}, \tag{4.55}$$

$$\langle jm'|J_-|jm \rangle = \hbar\sqrt{(j+m)(j-m+1)}\delta_{m',m-1}. \tag{4.56}$$

Next, it follows from the definition of the ladder operators given in (4.23) that we have:

$$J_1 = \frac{1}{2}(J_+ + J_-), \qquad J_2 = \frac{1}{2i}(J_+ - J_-). \tag{4.57}$$

From these relations it follows that if we know matrix representations of $J_+$ and $J_-$ then we can use them to determine the corresponding matrix representations of $J_1$ and $J_2$ (actually, this needs a "little" proof, but you should be able to convince yourself that it is true – operators and their matrix representations are different mathematical objects).

Now we can begin our construction of matrix representations of the angular momentum operators for different values of $j$.

$\boxed{\mathbf{j = 0}}$

This is an "easy" case. There is only one eigenvector, $|00\rangle$, and it follows from (4.53) – (4.56) that all the relevant matrix elements are zero. Therefore the matrix representations of $\mathbf{J}^2$, $J_1$, $J_2$, and $J_3$ are all zero.

$\boxed{\mathbf{j} = \tfrac{1}{2} \text{: The ``spin } \tfrac{1}{2} \text{ representation''}}$

In this case the relevant eigenvectors are $|\tfrac{1}{2}, \tfrac{1}{2}\rangle$ and $|\tfrac{1}{2}, -\tfrac{1}{2}\rangle$. The "general form" of the matrix representation is as follows:

$$\begin{pmatrix} \langle \tfrac{1}{2}, \tfrac{1}{2} | \ |\tfrac{1}{2}, \tfrac{1}{2}\rangle & \langle \tfrac{1}{2}, \tfrac{1}{2} | \ |\tfrac{1}{2}, -\tfrac{1}{2}\rangle \\ \\ \langle \tfrac{1}{2}, -\tfrac{1}{2} | \ |\tfrac{1}{2}, \tfrac{1}{2}\rangle & \langle \tfrac{1}{2}, -\tfrac{1}{2} | \ |\tfrac{1}{2}, -\tfrac{1}{2}\rangle \end{pmatrix}, \tag{4.58}$$

where by the phrase "general form" we mean that the matrix representation for a particular operator (e.g. $J_3$) is obtained by inserting the operator in the "gap" between the bra's and ket's in the general form, and using (4.53) – (4.56) to compute the associated matrix elements.

Therefore, the matrix representation for $J_3$ is given by:

$$\begin{pmatrix} \langle \tfrac{1}{2}, \tfrac{1}{2} | J_3 | \tfrac{1}{2}, \tfrac{1}{2}\rangle & \langle \tfrac{1}{2}, \tfrac{1}{2} | J_3 | \tfrac{1}{2}, -\tfrac{1}{2}\rangle \\ \\ \langle \tfrac{1}{2}, -\tfrac{1}{2} | J_3 | \tfrac{1}{2}, \tfrac{1}{2}\rangle & \langle \tfrac{1}{2}, -\tfrac{1}{2} | J_3 | \tfrac{1}{2}, -\tfrac{1}{2}\rangle \end{pmatrix} = \frac{\hbar}{2}\begin{pmatrix} 1 & 0 \\ 0 & -1 \end{pmatrix}. \tag{4.59}$$

The matrix representation for $J_-$ is given by:

$$\begin{pmatrix} \langle \tfrac{1}{2}, \tfrac{1}{2} | J_- | \tfrac{1}{2}, \tfrac{1}{2}\rangle & \langle \tfrac{1}{2}, \tfrac{1}{2} | J_- | \tfrac{1}{2}, -\tfrac{1}{2}\rangle \\ \\ \langle \tfrac{1}{2}, -\tfrac{1}{2} | J_- | \tfrac{1}{2}, \tfrac{1}{2}\rangle & \langle \tfrac{1}{2}, -\tfrac{1}{2} | J_- | \tfrac{1}{2}, -\tfrac{1}{2}\rangle \end{pmatrix} = \hbar \begin{pmatrix} 0 & 0 \\ 1 & 0 \end{pmatrix} \tag{4.60}$$

The matrix representation for $J_+$ is given by:

$$\begin{pmatrix} \langle \tfrac{1}{2}, \tfrac{1}{2} | J_+ | \tfrac{1}{2}, \tfrac{1}{2}\rangle & \langle \tfrac{1}{2}, \tfrac{1}{2} | J_+ | \tfrac{1}{2}, -\tfrac{1}{2}\rangle \\ \\ \langle \tfrac{1}{2}, -\tfrac{1}{2} | J_+ | \tfrac{1}{2}, \tfrac{1}{2}\rangle & \langle \tfrac{1}{2}, -\tfrac{1}{2} | J_+ | \tfrac{1}{2}, -\tfrac{1}{2}\rangle \end{pmatrix} = \hbar \begin{pmatrix} 0 & 1 \\ 0 & 0 \end{pmatrix} \tag{4.61}$$

Using the matrix representations for $J_-$ and $J_+$ with (4.57), the matrix representations for $J_1$ and $J_2$, respectively, are easily obtained and are found to be:

$$\frac{\hbar}{2}\begin{pmatrix} 0 & 1 \\ 1 & 0 \end{pmatrix}, \qquad \frac{\hbar}{2}\begin{pmatrix} 0 & -i \\ i & 0 \end{pmatrix} \tag{4.62}$$

The spin $\tfrac{1}{2}$ matrix representations for $J_1$, $J_1$, $J_3$ (neglecting the factor $\tfrac{\hbar}{2}$) occur frequently enough in many applications that they warrant a special name. They are referred to as the *Pauli spin matrices*, which are denoted as follows:

$$
X = \begin{pmatrix} 0 & 1 \\ 1 & 0 \end{pmatrix}, \qquad
Y = \begin{pmatrix} 0 & -i \\ i & 0 \end{pmatrix}, \qquad
Z = \begin{pmatrix} 1 & 0 \\ 0 & -1 \end{pmatrix}. \tag{4.63}
$$

$\boxed{\mathbf{j = 1}\text{: The ``spin 1 representation''}}$

We proceed exactly as in the case for $j = \frac{1}{2}$. In this case the relevant eigenvectors are $|1, 1\rangle$, $|1, 0\rangle$ and $|1, -1\rangle$. The "general form" of the matrix representation is as follows:

$$
\begin{pmatrix}
\langle 1, 1| \ \ |1, 1\rangle & \langle 1, 1| \ \ |1, 0\rangle & \langle 1, 1| \ \ |1, -1\rangle \\
\\
\langle 1, 0| \ \ |1, 1\rangle & \langle 1, 0| \ \ |1, 0\rangle & \langle 1, 0| \ \ |1, -1\rangle \\
\\
\langle 1, -1| \ \ |1, 1\rangle & \langle 1, -1| \ \ |1, 0\rangle & \langle 1, -1| \ \ |1, -1\rangle
\end{pmatrix} \tag{4.64}
$$

where by the phrase "general form", as above, we mean that the matrix representation for a particular operator (e.g. $J_3$) is obtained by inserting the operator in the "gap" between the bra's and ket's in the general form, and using (4.53) – (4.56) to compute the associated matrix elements.

Therefore, the matrix representation for $J_3$ is given by:

$$
\begin{pmatrix}
\langle 1, 1|J_3|1, 1\rangle & \langle 1, 1|J_3|1, 0\rangle & \langle 1, 1|J_3|1, -1\rangle \\
\\
\langle 1, 0|J_3|1, 1\rangle & \langle 1, 0|J_3|1, 0\rangle & \langle 1, 0|J_3|1, -1\rangle \\
\\
\langle 1, -1|J_3|1, 1\rangle & \langle 1, -1|J_3|1, 0\rangle & \langle 1, -1|J_3|1, -1\rangle
\end{pmatrix}
$$

$$
= \hbar \begin{pmatrix} 1 & 0 & 0 \\ 0 & 0 & 0 \\ 0 & 0 & -1 \end{pmatrix} \tag{4.65}
$$

The matrix representation for $J_-$ is given by:

$$\begin{pmatrix} \langle 1, 1|J_-|1, 1\rangle & \langle 1, 1|J_-|1, 0\rangle & \langle 1, 1|J_-|1, -1\rangle \\ \langle 1, 0|J_-|1, 1\rangle & \langle 1, 0|J_-|1, 0\rangle & \langle 1, 0|J_-|1, -1\rangle \\ \langle 1, -1|J_-|1, 1\rangle & \langle 1, -1|J_-|1, 0\rangle & \langle 1, -1|J_-|1, -1\rangle \end{pmatrix}$$

$$= \hbar \begin{pmatrix} 0 & 0 & 0 \\ \sqrt{2} & 0 & 0 \\ 0 & \sqrt{2} & 0 \end{pmatrix} \tag{4.66}$$

The matrix representation for $J_+$ is given by:

$$\begin{pmatrix} \langle 1, 1|J_+|1, 1\rangle & \langle 1, 1|J_+|1, 0\rangle & \langle 1, 1|J_+|1, -1\rangle \\ \langle 1, 0|J_+|1, 1\rangle & \langle 1, 0|J_+|1, 0\rangle & \langle 1, 0|J_+|1, -1\rangle \\ \langle 1, -1|J_+|1, 1\rangle & \langle 1, -1|J_+|1, 0\rangle & \langle 1, -1|J_+|1, -1\rangle \end{pmatrix}$$

$$= \hbar \begin{pmatrix} 0 & \sqrt{2} & 0 \\ 0 & 0 & \sqrt{2} \\ 0 & 0 & 0 \end{pmatrix} \tag{4.67}$$

Using the matrix representations for $J_-$ and $J_+$ with (4.57), the matrix representations for $J_1$ and $J_2$, respectively, are easily obtained.

## 4.4 Orbital Angular Momentum

Recall the way in which we began our study of the quantum theory of angular momentum. We began with the *classical* expression of orbital angular momentum of a particle about a point, and then "quantized" this expression by associating position and momentum operators to the position and momentum coordinates (Definition 4.13). Then the commutation relations for position and momentum operators that have been considered earlier in this book were used to derive commutation relations for the different components of the (orbital) angular momentum (Proposition 2). We then made the "leap" of defining a "general" quantum mechanical angular momentum operator as any three self-adjoint operators satisfying the commutation relations derived for the orbital angular momentum (Definition 4.18). The theory subsequently developed was solely a consequence of these three commutation relations for the three self-adjoint operators. We justified this leap into abstraction by saying that there we would encounter different types of angular momentum and, quantum mechanically, their common feature is the commutation relations. There is a justification for this statement, but it involves a deeper consideration of symmetries and rotations than we will go into in this book. An excellent discussion can be found in the "Littlejohn Notes" or in Cohen-Tannoudji et al.[3]

However, the important point we want to make now is the following. The allowed values of $j$ and $m$ that we obtained above follow solely from the commutation relations for $J_1$, $J_2$, and $J_3$ (convince your self of this). A particular "type" of angular momentum may have physical constraints that limit those values in some way. That is what we now want to address by returning to that particular type of angular momentum from which we started, orbital angular momentum. In particular, we have the following result.

**Theorem 11.** *For* orbital *angular momentum $j$ and $m$ must both be integers.*

**Proof.** We will give the standard "proof" of this result (and explain why we put "proof" in quotes afterwards).

Our work in this part of the book with angular momentum has been completely algebraic in nature. That is, we have not considered wave functions as you did earlier in the book. A natural "next step" in our study of angular momentum would be to consider angular momentum wave functions, i.e. spatial dependence of angular momentum. This topic is discussed extremely well in the "Littlejohn Notes". We will not pursue this topic here. However, for this "proof" we will need to consider an eigenfunction of $L_3$ (not $J_3$ since we are considering orbital angular momentum).

A point in space written in spherical coordinates has the form:

$$\mathbf{x} = (r \sin\theta \cos\phi, r \sin\theta \sin\phi, r \cos\theta) \tag{4.68}$$

---

[3]C. Cohen-Tannoudji, B. Diu, and F. Laloe. *Quantum Mechanics*. Wiley-VCH, 1992.

and using the chain rule we have:

$$\frac{\partial}{\partial\phi} = \frac{\partial x_1}{\partial\phi}\frac{\partial}{\partial x_1} + \frac{\partial x_2}{\partial\phi}\frac{\partial}{\partial x_2} + \frac{\partial x_3}{\partial\phi}\frac{\partial}{\partial x_3},$$

$$= -x_2\frac{\partial}{\partial x_1} + x_1\frac{\partial}{\partial x_2}. \tag{4.69}$$

From the last line of this expression (using $P_k = \frac{\hbar}{i}\frac{\partial}{\partial x_k}$, and associating the position operator $X_k$ with the position coordinate $x_k$, $k = 1, 2$), we see that:

$$-i\hbar\frac{\partial}{\partial\phi} = X_1 P_2 - P_1 X_2 = L_3 \tag{4.70}$$

Now let $\psi_m(r, \theta, \phi)$ denote an eigenfunction of $L_3$. Then we have:

$$-i\hbar\frac{\partial\psi_m}{\partial\phi} = L_3\psi_m = m\hbar\psi_m \tag{4.71}$$

or

$$\frac{\partial\psi_m}{\partial\phi} = im\psi_m. \tag{4.72}$$

Integrating this expression from 0 to $\phi$ (leaving $r$ and $\theta$ fixed) gives:

$$\psi_m(r, \theta, \phi) = e^{im\phi}\psi_m(r, \theta, 0). \tag{4.73}$$

Physically, we require the wavefunction to be a single valued function of the coordinates, i.e. at a given point in space, the wavefunction assumes only one value. In other words, we require:

$$\psi_m(r, \theta, \phi) = \psi_m(r, \theta, \phi + 2\pi) = e^{im(\phi+2\pi)}\psi_m(r, \theta, 0),$$

$$= e^{2\pi im}\left(e^{im\phi}\psi_m(r, \theta, 0)\right) = e^{2\pi im}\psi_m(r, \theta, \phi). \tag{4.74}$$

Hence, the only way that:

$$\psi_m(r, \theta, \phi) = e^{2\pi im}\psi_m(r, \theta, \phi), \tag{4.75}$$

is if $m$ is an integer, which implies that $j$ must be an integer (why?). $\qquad\square$

Now we need to explain why we put quotation marks around the word "proof". The result is certainly true, and there are no errors in our computations. The issue comes with the statement "Physically, we require the wavefunction to be a single valued function of the coordinates". Why should this be true? The wavefunction is not a "physical field", unlike pressure or density, where the value of such a field corresponds to a "physical observable, i.e. there is a unique value of the observable at each point in space. Recall our interpretation of the wavefunction as a *probability density.* Moreover, the wavefunction is "arbitrary" up to a phase factor, e.g a wavefunction $\psi$ should have the same physical implications as $-\psi$. Clearly, a deeper analysis of this notion of "single valuedness" of the wave is required.[4] A resolution comes in a deeper understanding of the nature of the eigenfunctions of the Schrödinger equation (e.g. the Hilbert space of its solutions), which is beyond the scope of this book. However, a proof that the eigenvalues of the orbital angular momentum operator must take on integer values can be found in the paper of Kaplan and Yu.[5] The proof is surprisingly simple (at the level of the mathematics that we have already done in the book) and it uses the algebraic approach of operators that we have been developing, rather than wave mechanics, which was the focus of the first part of the book. However, as we have mentioned, a natural further development of angular momentum would be to develop angular momentum eigenfunctions that are appropriate solutions of the Schrödinger equation. This is pursued in the "Littlejohn Notes" as well as in Cohen-Tannoudji et al.[6]

### *Spin, and the Stern-Gerlach Experiment*

We have mentioned that there are types of angular momentum other than orbital angular momentum, and *spin* is probably the preeminent example. Despite its name, spin is an intrinsically quantum mechanical property with no classical analog. It was definitively demonstrated in the 1922 experiment of Otto Stern and Walther Gerlach (the "Stern-Gerlach experiment"). The background and history are well described in the article of Friedrich and Herschbach.[7]

The following two quotes taken from this article describe its importance in the history of physics, as well as its continuing broad impact.

*The demonstration of space quantization, carried out in Frankfurt, Germany, in 1922 by Otto Stern and Walther Gerlach, ranks among the dozen or so canonical experiments that ushered in the heroic age of quantum physics.*

---

[4]The book by Schumacher and Westmoreland has a nice section call "How not to think about $\psi$", (section 10.6) which we recommend.

B. Schumacher and M. Westmoreland. *Quantum processes systems, and information.* Cambridge University Press, 2010.

[5]David M. Kaplan and F.Y. Wu. On the eigenvalues of orbital angular momentum. *Chinese Journal of Physics*, 9(1):31–33, 1971.

[6]C. Cohen-Tannoudji, B. Diu, and F. Laloe. *Quantum Mechanics.* Wiley-VCH, 1992.

[7]B. Friedrich and D. Herschbach. Stern and Gerlach: How a bad cigar helped reorient atomic physics. *Physics Today*, 56(12):53–59, 2003.

*Descendants of the Stern-Gerlach experiment (SGE) and its key concept of sorting quantum states via space quantization are legion. Among them are the prototypes for nuclear magnetic resonance, optical pumping, the laser, and atomic clocks, as well as incisive discoveries such as the Lamb shift and the anomalous increment in the magnetic moment of the electron, which launched quantum electrodynamics The means to probe nuclei, proteins, and galaxies; image bodies and brains; perform eye surgery, read music or data from compact discs; and scan bar codes on grocery packages or DNA base pairs in the human genome all stem from exploiting transitions between space-quantized quantum states.*

In the next chapter the general framework of the Stern-Gerlach experiment will play a a central role in several ways. Most notably, in the topic of "measurement" in quantum mechanics. Now we want to explain the essentials of this experiment (which is no substitute for a detailed study of the underlying physics – a good discussion can be found in Cohen-Tannoudji et al.,[8] Bohm,[9] or Rodriguez et al.[10]).

Stern and Gerlach sent a beam of neutral atoms through a region of an inhomogeneous magnetic field. This means that the magnetic field had a nonzero spatial gradient, which was chosen to be the $z$ direction. Stern and Gerlach used silver atoms in their experiment (Ag; Z = 47). This was a very deliberate choice. We will briefly explain the reasons behind this choice (you don't need to know these details – the main point is at the end of our discussion). In the ground state a silver atom has one *valence electron* in a 5s subshell. The other 46 electrons fill all subshells for n = 1, n = 2, and n = 3, and the 4d subshell. *All these closed shells and subshells contribute zero to the total angular momentum of the atom. So the properties measured in the Stern-Gerlach experiment are the properties of the valence electron.* This is very significant. Effectively, for the properties under consideration, the Stern-Gerlach experiment is an experiment on single electrons, and the angular momentum of the electron in the 5s subshell corresponds to $l = 0$.

---

[8]C. Cohen-Tannoudji, B. Diu, and F. Laloe. *Quantum Mechanics.* Wiley-VCH, 1992.
[9]D. Bohm. *Quantum theory.* Courier Corporation, 1951.
[10]E. Benitez Rodriguez, L.M. Arevalo Aguilar, and E. Piceno Martinez. A full quantum analysis of the Stern – Gerlach experiment using the evolution operator method: Analyzing current issues in teaching quantum mechanics. *European Journal of Physics*, 38(2):025403, 2017.

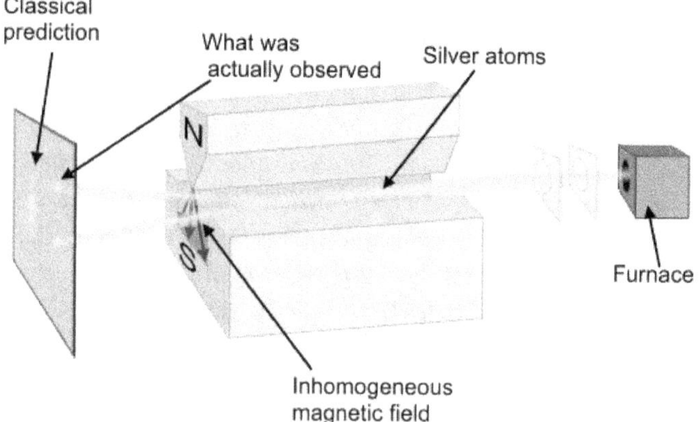



Figure 4.1: Schematic of the Stern-Gerlach experiment (figure from Wikipedia).

The following is "what happens": When a silver atom passes through the magnetic field it experiences a deflection (as a result of the magnetic field gradient) depending on the angular momentum of the particle. If the silver atom behaved as a classical particle, it is expected that the angular momentum vectors of the beam of particles would be randomly distributed, and the beam would strike a detector in a "broad spot" centered around the axis of the beam (see Figure 4.1 for an illustration). This was not observed. The silver atoms did not behave as classical particles. The silver atoms had total (quantum) orbital angular momentum zero (with the electrons occupying the various shells and subhells as described above). Nevertheless, the beam struck the detector in two discrete locations, and it was a big mystery as to why. (see Figure 4.1 for an illustration of what was observed). It could be argued that deflection of the beam of silver atoms due to the force generated by the inhomogeneous field was due to the interaction of the magnetic field with the "angular momentum" of the silver atoms, which was completely due to the angular momentum of the valence electron, which corresponded to $l = 0$ for the electron in the 5s subshell. The distribution of silver atoms on the detector could not be explained by classical angular momentum, and it could not be explained by quantum orbital angular momentum. So what was going on?

The paper of Friedrich and Herschbach[11] describes the efforts to come up with an explanation. Goudsmit and Uhlenbeck finally proposed the idea

---

[11]B. Friedrich and D. Herschbach. Stern and Gerlach: How a bad cigar helped reorient atomic physics. *Physics Today*, 56(12):53–59, 2003.

of an "intrinsic angular momentum" (or "spin") that provided an explanation of the results.

If you would like to read Goudsmit's own personal account of the discovery of electron spin go to the following URL:

http://www.lorentz.leidenuniv.nl/history/spin/goudsmit.html

A very complete account of the topic of spin is given in the book of Tomonaga and Oka.[12]

For the purpose of the rest of the book, the important facts are that the spin of an electron is described by a "two state quantum mechanical system" – spin "up" and spin "down". where up and down are defined by the direction of the gradient of the magnetic field. In this case the gradient is in the $z$ direction. We take spin up as the positive $z$ direction and spin down as the negative $z$ direction (now we are in danger of beginning to encounter some conceptual difficulties by "confusing" the physical space with the abstract Hilbert space describing the "spin dynamics" of the electron – but we will address this later on). The "Stern-Gerlach device" sorts the beam of silver atoms into these two quantum "spin states". Next we want to describe such two state quantum systems in more detail, and that will take us into the next chapter.

## Problems

1. *Orbital angular momentum commutation relations.*
   Show that $[AB, C] = A[B, C] + [A, C]B$ for three operators $A, B, C$. Work out a similar expression for $[AB, CD]$. Hence show that the orbital angular momentum operators $L_1, L_2, L_3$, where $L_1 = X_2 P_3 - X_3 P_2$, $L_2 = X_3 P_1 - X_1 P_3$ and $L_3 = X_1 P_2 - X_2 P_1$, satisfy the angular momentum commutation relations

$$[L_j, L_k] = i\hbar \sum_{m=1}^{3} \epsilon_{jkm} L_m.$$

   ($X_j$ and $P_k$ satisfy the usual canonical commutation relations $[X_j, P_k] = i\hbar\delta_{jk}$.)

---

[12]Sin-Itiro Tomonaga and Takeshi Oka. *The Story of Spin*. University of Chicago Press Chicago, IL, 1997.

2. *Raising and lowering operators.*

   Consider three operators $J_1, J_2, J_3$ satisfying the angular momentum commutation relations. With the definitions $J_\pm = J_1 \pm i J_2$ and $J^2 = J_1^2 + J_2^2 + J_3^2$, show that

   $$[J^2, J_\pm] = 0;$$

   $$J_+ J_- = J^2 - J_3^2 + \hbar J_3;$$

   $$J_- J_+ = J^2 - J_3^2 - \hbar J_3;$$

   $$[J_+, J_-] = 2\hbar J_3;$$

   $$[J_3, J_\pm] = \pm\hbar J_\pm.$$

3. *Pauli matrices.*

   The Pauli matrices are defined by

   $$\sigma_x = \begin{pmatrix} 0 & 1 \\ 1 & 0 \end{pmatrix}; \quad \sigma_y = \begin{pmatrix} 0 & -i \\ i & 0 \end{pmatrix}; \quad \sigma_z = \begin{pmatrix} 1 & 0 \\ 0 & -1 \end{pmatrix}.$$

   (a) Show that

   $$\sigma_x^2 = I; \quad \sigma_y^2 = I; \quad \sigma_z \sigma_x = i\sigma_y; \quad \sigma_z \sigma_y = -i\sigma_x;$$

   where $I$ is the identity matrix on $\mathbb{C}^2$

   (b) Consider two 3-component vectors $\mathbf{A}$ and $\mathbf{B}$ with real entries. Let $\mathbf{A}.\sigma$ and $\mathbf{B}.\sigma$ be the matrices

   $$\mathbf{A}.\sigma = A_x \sigma_x + A_y \sigma_y + A_z \sigma_z; \quad \mathbf{B}.\sigma = B_x \sigma_x + B_y \sigma_y + B_z \sigma_z.$$

   Write down the matrices $\mathbf{A}.\sigma$ and $\mathbf{B}.\sigma$ explicitly, and hence, or otherwise show that

   $$(\mathbf{A}.\sigma)\,(\mathbf{B}.\sigma) = (\mathbf{A}.\mathbf{B})\,I + i(\mathbf{A} \times \mathbf{B}).\sigma.$$

   (c) Let $\mathbf{n}$ be a three component unit vector and let $a$ be a real parameter. Show that

   $$\exp\,(ia\mathbf{n}.\sigma) = I \cos a + i\mathbf{n}.\sigma \sin a$$

4. *The spin one representation.*

   For the $j = 1$ representation of the angular momentum commutation relations, write down the action of the operators $J_\pm$ and $J_3$ on basis vectors which are simultaneous eigenvectors of $J^2$ and $J_3$.

   Calculate the action of $J_1$ and $J_2$ in this case and hence show that the operators $J_1, J_2, J_3$ satisfy the angular momentum commutation relations $[J_j, J_k] = i\hbar \sum_{m=1}^{3} \epsilon_{jkm} J_m$.

Calculate the matrices of $J_1, J_2, J_3$ with respect to the basis you have used above, and show that these matrices also satisfy the angular momentum commutation relations.

5. *Matrix elements for angular momentum operators.*
   Let $J_1, J_2, J_3$ satisfy angular momentum commutation relations, and let $|j, m\rangle$ be the usual (normalised) simultaneous eigenvectors of $J^2$ and $J_3$ with eigenvalues $\hbar^2 j(j+1)$ and $m\hbar$ respectively.
   By considering $\langle j, m| J_+^2 |j, m\rangle$ or otherwise, show that

$$\langle j, m| J_1^2 |j, m\rangle = \langle j, m| J_2^2 |j, m\rangle,$$

and find the value of this matrix element.

6. *A spin Hamiltonian.*
   Consider the Hamiltonian

$$H = \frac{1}{2I}(J_1^2 + J_3^2),$$

where $I$ is a constant.

(a) Show that

$$\langle 1, m|H|1, m\rangle = \frac{1}{4I}(2 + m^2)\hbar^2,$$

   for $m = -1, 0, 1$.

(b) Show also that for $m \neq n$, $\langle 1, m|H|1, n\rangle = 0$ unless $(m, n) = (1, -1)$ or $(m, n) = (-1, 1)$. Calculate $\langle 1, 1|H|1, -1\rangle$.

(c) Deduce the three eigenvalues of $H$ and the corresponding eigenstates.

7. Let $J_1, J_2, J_3$ be self-adjoint angular momentum operators obeying:

$$[J_1, J_2] = i\hbar J_3, \quad [J_2, J_3] = i\hbar J_1, \quad [J_3, J_1] = i\hbar J_2,$$

with the operators $\mathbf{J}^2$, $J_+$ and $J_-$ defined by $\mathbf{J}^2 = J_1^2 + J_2^2 + J_3^2$ and $J_\pm = J_1 \pm iJ_2$. We denote a simultaneous eigenvector of $\mathbf{J}^2$ and $J_3$ by $|jm\rangle$, i.e.

$$\mathbf{J}^2|jm\rangle = \hbar^2 j(j+1)|jm\rangle,$$

$$J_3|jm\rangle = \hbar m|jm\rangle.$$

Consider the Hamiltonian:

$$H = \mathbf{J}^2 - J_3^2. \tag{4.76}$$

(a) Show that $H$ is self-adjoint.

(b) Show that

$$\mathbf{J}^2 = J_+ J_- + J_3^2 - \hbar J_3 = J_- J_+ + J_3^2 + \hbar J_3.$$

(c) Compute the eigenvalues of $H$.

(d) Are $|jm\rangle$ eigenvectors for $H$ *and* $\mathbf{J}^2$? Justify your answer.

(e) Suppose the state space for $H$ is $C^2$, with basis vectors $\{|\frac{1}{2}, \frac{1}{2}\rangle, |\frac{1}{2}, -\frac{1}{2}\rangle\}$. Compute the matrix representation of $H$ with respect to this basis.

8. Let $\mathbf{J} = (J_1, J_2, J_3)$ where $J_1$, $J_2$, $J_3$ are self-adjoint angular momentum operators obeying:

$$[J_1, J_2] = i\hbar J_3, \quad [J_2, J_3] = i\hbar J_1, \quad [J_3, J_1] = i\hbar J_2,$$

with the operators $\mathbf{J}^2$, $J_+$ and $J_-$ defined by $\mathbf{J}^2 = J_1^2 + J_2^2 + J_3^2$ and $J_\pm = J_1 \pm iJ_2$. We denote a simultaneous eigenvector of $\mathbf{J}^2$ and $J_3$ by $|jm\rangle$, i.e.

$$\mathbf{J}^2 |jm\rangle = \hbar^2 j(j+1)|jm\rangle,$$

$$J_3|jm\rangle = \hbar m|jm\rangle.$$

Consider the Hamiltonian:

$$H = \frac{1}{2I_1}\left(J_1^2 + J_2^2\right) + \frac{1}{2I_3}J_3^2,$$

where $I_1$ and $I_3$ are positive constants.

(a) Show that $\mathbf{J} \times \mathbf{J} = i\hbar\mathbf{J}$.

(b) Show that

$$\mathbf{J}^2 = J_+ J_- + J_3^2 - \hbar J_3 = J_- J_+ + J_3^2 + \hbar J_3.$$

(c) Using the simultaneous eigenvectors of $\mathbf{J}^2$ and $J_3$, compute the corresponding eigenvalues of $H$.

(d) Suppose the state space for $H$ is $\mathbb{C}^2$, with basis vectors $\{|\frac{1}{2}, \frac{1}{2}\rangle, |\frac{1}{2}, -\frac{1}{2}\rangle\}$. Compute the matrix representation of $H$ with respect to this basis.

(e) Compute the expectation value of $J_3^2$ in the state $|\frac{1}{2}, -\frac{1}{2}\rangle$.

9. Consider the following three matrices and their respective eigenvalues and normalized eigenvectors:

$$X = \begin{pmatrix} 0 & 1 \\ 1 & 0 \end{pmatrix}, \quad +1, \; \frac{1}{\sqrt{2}}\begin{pmatrix} 1 \\ 1 \end{pmatrix}, \quad -1, \; \frac{1}{\sqrt{2}}\begin{pmatrix} 1 \\ -1 \end{pmatrix}.$$

$$Y = \begin{pmatrix} 0 & -i \\ i & 0 \end{pmatrix}, \quad +1, \; \frac{1}{\sqrt{2}}\begin{pmatrix} 1 \\ i \end{pmatrix}, \quad -1, \; \frac{1}{\sqrt{2}}\begin{pmatrix} 1 \\ -i \end{pmatrix}.$$

$$Z = \begin{pmatrix} 1 & 0 \\ 0 & -1 \end{pmatrix}, \quad +1, \; \begin{pmatrix} 1 \\ 0 \end{pmatrix}, \quad -1, \; \begin{pmatrix} 0 \\ 1 \end{pmatrix}.$$

We define

$$S^2 \equiv X^2 + Y^2 + Z^2,$$

and

$$S^+ \equiv X + iY, \qquad S^- \equiv X - iY.$$

(a) Show that $S^2$ commutes with $X$, $Y$, and $Z$.

(b) Are $S^+$ and $S^-$ self-adjoint? (Justify your answer.)

(c) Show that $S^+$ acting on $\begin{pmatrix} 0 \\ 1 \end{pmatrix}$ is an eigenvector of $Z$. Similarly, show that $S^-$ acting on $\begin{pmatrix} 1 \\ 0 \end{pmatrix}$ is an eigenvector of $Z$.

(d) Find two normalized vectors that are simultaneously eigenvectors of $S^2$ and $Y$. (Justify your answer.)

# Chapter 5

# Composite Systems, Tensor Products, and Entanglement

We will now describe some of the aspects of quantum mechanics that are most at odds with our intuition from classical mechanics. These topics are of intense current interest and, potentially, will be at the heart of significant technological advances (the phrase "quantum engineering" is becoming more and more common).[1] Our entry into this area of quantum mechanics will be through the consideration of an idealized Stern-Gerlach experiment. A deeper consideration of the full nature of the Stern-Gerlach experiment can be found in the paper of Rodriguez et al.[2]

## 5.1  The Stern-Gerlach Experiment–Revisited

We return to the Stern-Gerlach experiment in order to study more deeply issues associated with quantum measurement. In particular, we will consider quantum particles having spin $\frac{1}{2}$, i.e. particles whose behavior is described by quantum mechanics, such as electrons.

---

[1] Jonathan P Dowling and Gerard J Milburn. Quantum technology: the second quantum revolution. *Philosophical Transactions of the Royal Society of London. Series A: Mathematical, Physical and Engineering Sciences*, 361(1809):1655–1674, 2003.

Lars Jaeger. The second quantum revolution. *From Entanglement to Quantum Computing and Other Super-Technologies. Copernicus*, 2018.

[2] E. Benitez Rodriguez, L.M. Arevalo Aguilar, and E. Piceno Martinez. A full quantum analysis of the Stern–Gerlach experiment using the evolution operator method: Analyzing current issues in teaching quantum mechanics. *European Journal of Physics*, 38(2): 025403, 2017.

### The Pauli Spin Matrices and Dirac Notation–Review of Some Background

We begin by first developing some necessary mathematical tools.

### The Complex Vector Space, $\mathbb{C}^2$

When a beam of electrons[3] is passed through a Stern–Gerlach apparatus the beam is split into two–half correspond to electrons whose spin is in the direction of the magnetic field gradient $\left(\text{spin } \frac{1}{2}\right)$ and the other correspond to electrons whose spin is in the opposite direction of the magnetic field gradient $\left(\text{spin } - \frac{1}{2}\right)$. If we characterize the electrons solely by their spin, then this system provides an ideal example of a "two state system". The Hilbert space is two dimensional, and can be taken as $\mathbb{C}^2$, which we view as a (complex) linear vector space equipped with an inner product.

### The Pauli Spin Matrices

Recall the Pauli spin matrices that we derived in the previous chapter:

$$J_1 = \frac{\hbar}{2}X, \quad J_2 = \frac{\hbar}{2}Y, \quad J_3 = \frac{\hbar}{2}Z, \tag{5.1}$$

where

$$X = \begin{pmatrix} 0 & 1 \\ 1 & 0 \end{pmatrix}, \quad Y = \begin{pmatrix} 0 & -i \\ i & 0 \end{pmatrix}, \quad Z = \begin{pmatrix} 1 & 0 \\ 0 & -1 \end{pmatrix}. \tag{5.2}$$

We can now state the problem that we will address.

**Statement of Problem.** *We imagine a beam of spin $\frac{1}{2}$ quantum particles propagating in the y direction and passing through certain Stern–Gerlach devices. We will be considering measurements of $J_3$ and $J_1$. The relevant Hilbert space is $\mathbb{C}^2$, and $J_3$ and $J_1$ are Hermitian operators defined on this Hilbert space.*

We first need to develop some mathematical properties of $J_3$ and $J_1$ that will be necessary in applying the postulates of quantum mechanics for the purpose of measurement.

---

[3]Remember, Stern and Gerlach did not use a beam of electrons, but a beam of silver atoms. However, recall the discussion. The silver atoms were such that the measure properties were those of the single valence electron. We will abuse history, slightly, by referring to a "beam of electrons".

$$J_3 = \frac{\hbar}{2} \begin{pmatrix} 1 & 0 \\ 0 & -1 \end{pmatrix} = \frac{\hbar}{2} Z$$

The eigenvalues of $Z$ are 1, with corresponding eigenvector $\begin{pmatrix} 1 \\ 0 \end{pmatrix}$, and $-1$, with corresponding eigenvector $\begin{pmatrix} 0 \\ 1 \end{pmatrix}$. We will use Dirac's bra-ket notation in our calculations. Different notations are used in the literature for the ket vector corresponding to $\begin{pmatrix} 1 \\ 0 \end{pmatrix}$. Some of these are collected below.

$$\begin{pmatrix} 1 \\ 0 \end{pmatrix} : \quad |\uparrow\rangle, \ |0\rangle, \ |+\tfrac{1}{2}\rangle, \ |+z\rangle. \tag{5.3}$$

Similarly, notations for the ket vector corresponding to $\begin{pmatrix} 0 \\ 1 \end{pmatrix}$ are:

$$\begin{pmatrix} 0 \\ 1 \end{pmatrix} : \quad |\downarrow\rangle, \ |1\rangle, \ |-\tfrac{1}{2}\rangle, \ |-z\rangle. \tag{5.4}$$

We will use the notation $|\uparrow\rangle$ for $\begin{pmatrix} 1 \\ 0 \end{pmatrix}$ and $|\downarrow\rangle$ for $\begin{pmatrix} 0 \\ 1 \end{pmatrix}$ in this section. The projection operator onto the eigenvector $|\uparrow\rangle$ is given by:

$$|\uparrow\rangle\langle\uparrow| = \begin{pmatrix} 1 \\ 0 \end{pmatrix} (1 \ \ 0) = \begin{pmatrix} 1 & 0 \\ 0 & 0 \end{pmatrix}. \tag{5.5}$$

Similarly, the projection operator onto the eigenvector $|\downarrow\rangle$ is given by:

$$|\downarrow\rangle\langle\downarrow| = \begin{pmatrix} 0 \\ 1 \end{pmatrix} (0 \ \ 1) = \begin{pmatrix} 0 & 0 \\ 0 & 1 \end{pmatrix}. \tag{5.6}$$

Consequently, it is easy to see that the spectral decomposition for $J_3$ is given by:

$$J_3 = \frac{\hbar}{2}|\uparrow\rangle\langle\uparrow| - \frac{\hbar}{2}|\downarrow\rangle\langle\downarrow| = \frac{\hbar}{2} \begin{pmatrix} 1 & 0 \\ 0 & -1 \end{pmatrix} \tag{5.7}$$

$$J_1 = \frac{\hbar}{2} \begin{pmatrix} 0 & 1 \\ 1 & 0 \end{pmatrix} = \frac{\hbar}{2} X$$

The eigenvalues of $X$ are 1, with corresponding eigenvector $\frac{1}{\sqrt{2}}\begin{pmatrix} 1 \\ 1 \end{pmatrix}$, and $-1$, with corresponding eigenvector $\frac{1}{\sqrt{2}}\begin{pmatrix} 1 \\ -1 \end{pmatrix}$. Some notation for kets corresponding to $\frac{1}{\sqrt{2}}\begin{pmatrix} 1 \\ 1 \end{pmatrix}$ and $\frac{1}{\sqrt{2}}\begin{pmatrix} 1 \\ -1 \end{pmatrix}$, respectively, are given below:

$$\frac{1}{\sqrt{2}}\begin{pmatrix} 1 \\ 1 \end{pmatrix}: \quad |\rightarrow\rangle, |+\rangle, |+x\rangle, \tag{5.8}$$

$$\frac{1}{\sqrt{2}}\begin{pmatrix} 1 \\ -1 \end{pmatrix}: \quad |\leftarrow\rangle, |-\rangle, |-x\rangle. \tag{5.9}$$

The corresponding projection operators onto the respective eigenvectors are given by:

$$|\rightarrow\rangle\langle\rightarrow| = \frac{1}{2}\begin{pmatrix} 1 \\ 1 \end{pmatrix}(1 \; 1) = \frac{1}{2}\begin{pmatrix} 1 & 1 \\ 1 & 1 \end{pmatrix}, \tag{5.10}$$

$$|\leftarrow\rangle\langle\leftarrow| = \frac{1}{2}\begin{pmatrix} 1 \\ -1 \end{pmatrix}(1 \; -1) = \frac{1}{2}\begin{pmatrix} 1 & -1 \\ -1 & 1 \end{pmatrix}, \tag{5.11}$$

and the spectral decomposition of $J_1$ has the following form:

$$J_1 = \frac{\hbar}{2}|\rightarrow\rangle\langle\rightarrow| - \frac{\hbar}{2}|\leftarrow\rangle\langle\leftarrow| = \frac{\hbar}{2}\begin{pmatrix} 0 & 1 \\ 1 & 0 \end{pmatrix}. \tag{5.12}$$

*Change of Basis.* We will need to relate the basis of eigenvectors of $J_3$ to the basis of eigenvectors of $J_1$:

$$\begin{pmatrix} 1 \\ 0 \end{pmatrix}, \begin{pmatrix} 0 \\ 1 \end{pmatrix} \leftrightarrow \frac{1}{\sqrt{2}}\begin{pmatrix} 1 \\ 1 \end{pmatrix}, \frac{1}{\sqrt{2}}\begin{pmatrix} 1 \\ -1 \end{pmatrix}. \tag{5.13}$$

The relations between the two bases of eigenvectors can be determined essentially by inspection:

$$\begin{pmatrix} 1 \\ 0 \end{pmatrix} = \frac{1}{\sqrt{2}}\left( \frac{1}{\sqrt{2}}\begin{pmatrix} 1 \\ 1 \end{pmatrix} + \frac{1}{\sqrt{2}}\begin{pmatrix} 1 \\ -1 \end{pmatrix} \right) \tag{5.14}$$

$$\begin{pmatrix} 0 \\ 1 \end{pmatrix} = \frac{1}{\sqrt{2}}\left( \frac{1}{\sqrt{2}}\begin{pmatrix} 1 \\ 1 \end{pmatrix} - \frac{1}{\sqrt{2}}\begin{pmatrix} 1 \\ -1 \end{pmatrix} \right) \tag{5.15}$$

or, in bra-ket notation:

$$|\uparrow\rangle = \frac{1}{\sqrt{2}}|\rightarrow\rangle + \frac{1}{\sqrt{2}}|\leftarrow\rangle \qquad (5.16)$$

$$|\downarrow\rangle = \frac{1}{\sqrt{2}}|\rightarrow\rangle - \frac{1}{\sqrt{2}}|\leftarrow\rangle, \qquad (5.17)$$

and

$$|\rightarrow\rangle = \frac{1}{\sqrt{2}}|\uparrow\rangle + \frac{1}{\sqrt{2}}|\downarrow\rangle$$

$$|\leftarrow\rangle = \frac{1}{\sqrt{2}}|\uparrow\rangle - \frac{1}{\sqrt{2}}|\downarrow\rangle. \qquad (5.18)$$

## 5.2 Four "Thought Experiments" on the Measurement of Spin

Now we will consider four "thought experiments"[4] that illustrate measuring an observable in a quantum system. The observables will be $J_3$ and $J_1$ (recall that we have proven that these are Hermitian operators) and the Hilbert space defining the system is $\mathbb{C}^2$. In the "thought experiments" by the phrase "Stern-Gerlach apparatus" we will mean an apparatus having an inhomogeneous magnetic field whose gradient is in a specific direction. For our purposes the magnetic field gradient will either be in the vertical ($z$) direction or the horizontal ($x$ direction), and these directions are "local" to the particular Stern-Gerlach apparatus. A beam of spin $\frac{1}{2}$ particles (e.g. electrons) propagates in the $y$ direction (relative to the Stern-Gerlach apparatus) and passes through the inhomogeneous magnetic field associated with the Stern-Gerlach apparatuses in each experiment.

---

[4]The state vector $|\psi\rangle$ is a description of the beam of spin $\frac{1}{2}$ quantum particles. It is not the state vector for each individual particle in the beam (you might think about why I say this). The real question would be how does one realize a beam in this state in a "real experiment"? This is why we are using the phrase "thought experiment". It is mainly to give experience with the postulates of quantum mechanics, and quantum measurement. "Thought experiments" have played (and still play) an important role in quantum mechanics. We will see that later when we consider the Einstein-Podolsky-Rosen (EPR) paradox. It took quite a few years before the conditions of the "thought experiment" of EPR could be realized in the laboratory. However, the consequences of the thought experiment were so compelling that the need to design a "real experiment" was clear.

$$\boxed{\textbf{Experiment 1:}}$$

This is the "standard" Stern-Gerlach experiment. A beam of spin $\frac{1}{2}$ particles passes through a Stern-Gerlach apparatus with the inhomogeneous magnetic field gradient in the $z$ direction, see Fig. 5.1. We assume that the initial state of the beam of spin $\frac{1}{2}$ particles is given by:

$$|\psi\rangle = \alpha\,|\uparrow\rangle + \beta\,|\downarrow\rangle, \tag{5.19}$$

where normalization of $|\psi\rangle$ implies that $|\alpha|^2 + |\beta|^2 = 1$.

We wish to measure $J_3$. The possible values of $J_3$ that we can measure are its eigenvalues, $\pm\frac{\hbar}{2}$. Recalling 3.1, the probability that we measure $\frac{\hbar}{2}$ is given by:

$$\begin{aligned}
p\left(\frac{\hbar}{2}\right) &= \langle\psi|\,(|\uparrow\rangle\langle\uparrow|)\,|\psi\rangle, \\
&= (\alpha^*\langle\uparrow| + \beta^*\langle\downarrow|)\,(|\uparrow\rangle\langle\uparrow|)\,(\alpha\,|\uparrow\rangle + \beta\,|\downarrow\rangle), \\
&= |\alpha|^2.
\end{aligned} \tag{5.20}$$

A similar calculation shows that the probability that we measure $-\frac{\hbar}{2}$ is given by:

$$p\left(-\frac{\hbar}{2}\right) = |\beta|^2. \tag{5.21}$$

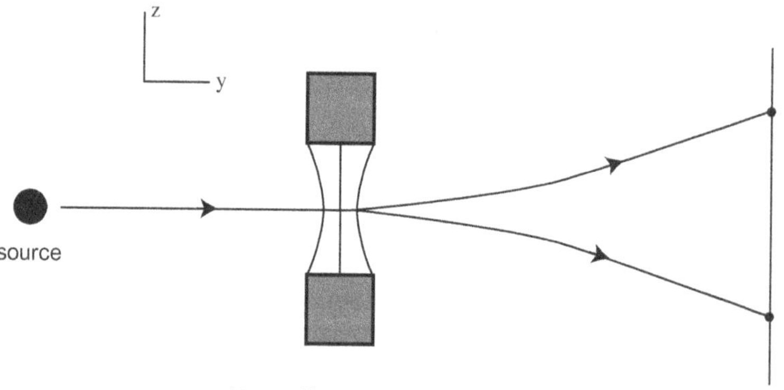


Figure 5.1: A beam of spin $\frac{1}{2}$ particles passing through a Stern-Gerlach apparatus, where the gradient of the magnetic field is in the $z$ direction.

So, as expected, the Stern-Gerlach apparatus splits the beam into two beams–one beam consisting of "spin up" particles and the other beam consisting of "spin down" particles (where "up" and "down" are measured with respect to the vertical ($z$) direction).

**Experiment 2:**

In this experiment the beam of spin $\frac{1}{2}$ particles passes through two consecutive Stern-Gerlach apparatuses, with each having their magnetic field gradient oriented in the $z$ direction. As in the previous experiment, the initial state (before the beam passes through either Stern-Gerlach apparatus) is given by:

$$|\psi\rangle = \alpha\,|\uparrow\rangle + \beta\,|\downarrow\rangle, \quad |\alpha|^2 + |\beta|^2 = 1. \tag{5.22}$$

We assume that a measurement has been made immediately after the beam passes through the first Stern-Gerlach apparatus and only particles with spin $J_3$ eigenvalues of $\frac{\hbar}{2}$ are allowed to continue to the second Stern-Gerlach apparatus. Recalling 3.2, the state following the measurement that gives the eigenvalue $\frac{\hbar}{2}$ is given by:

$$\frac{|\uparrow\rangle\langle\uparrow|\,(\alpha\,|\uparrow\rangle + \beta\,|\downarrow\rangle)}{\sqrt{\langle\psi|\,(|\uparrow\rangle\langle\uparrow|)\,|\psi\rangle}} = |\uparrow\rangle \tag{5.23}$$

This is consistent with the measurement postulate of quantum mechanics. The eigenvalue $\frac{\hbar}{2}$ is measured and the state "collapses" to the eigenstate of corresponding to the eigenvalue $\frac{\hbar}{2}$. The beam of particles in state $|\uparrow\rangle$ then passes through the second Stern-Gerlach apparatus. Another measurement of $J_3$ on is state is carried out, and the probabilities that $\frac{\hbar}{2}$ and $-\frac{\hbar}{2}$ are measured are, respectively, given by:

$$p\left(\frac{\hbar}{2}\right) = \langle\uparrow|\,(|\uparrow\rangle\langle\uparrow|)\,|\uparrow\rangle = 1,$$

$$p\left(-\frac{\hbar}{2}\right) = \langle\uparrow|\,(|\downarrow\rangle\langle\downarrow|)\,|\uparrow\rangle = 0. \tag{5.24}$$

Hence, we always get the outcome $\frac{\hbar}{2}$, as expected, since the state of the beam that enters the second Stern-Gerlach device is $|\uparrow\rangle$.

**Experiment 3:**

We consider a beam of spin $\frac{1}{2}$ particles in the initial state:

$$\psi = \alpha\,|\uparrow\rangle + \beta\,|\downarrow\rangle, \quad |\alpha|^2 + |\beta|^2 = 1. \tag{5.25}$$

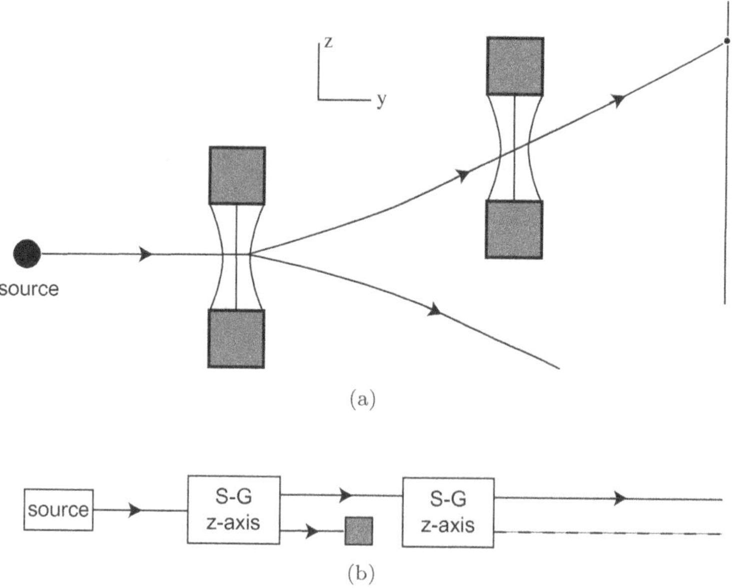

(a)

(b)

Figure 5.2: A beam of spin $\frac{1}{2}$ particles passes through two consecutive Stern-Gerlach devices, with each measuring $J_3$. a) After passing through the first Stern-Gerlach device only particles in the state $|\uparrow\rangle$ are allowed to pass through the second Stern-Gerlach device. b) A more "schematic" depiction of the experiment that is shown in a). Fig. 5.2b (from Wikipedia).

The beam passes through a Stern-Gerlach apparatus where $J_3$ is measured, and particles where the outcome of the measurement is $\frac{\hbar}{2}$ are allowed to pass through a second Stern-Gerlach device where the magnetic field gradient is in the $x$ direction. The particles passing through the second Stern-Gerlach device are in the state $|\uparrow\rangle$, and we wish to measure $J_1$. We know that the possible outcomes of this measurement are $\frac{\hbar}{2}$ and $-\frac{\hbar}{2}$ (the two eigenvalues of $J_1$). The probability that the outcome of the measurement is $\frac{\hbar}{2}$ is given by:

$$p\left(\frac{\hbar}{2}\right) = \langle\uparrow|\left(|\rightarrow\rangle\langle\rightarrow|\right)|\uparrow\rangle. \tag{5.26}$$

Now in order to calculate this quantity we need to express the state $|\uparrow\rangle$ in terms of the basis of $J_1$ given by $|\rightarrow\rangle$ and $|\leftarrow\rangle$. The formula relating the basis elements was previously given in (5.17), which we rewrite here:

$$|\uparrow\rangle = \frac{1}{\sqrt{2}}|\rightarrow\rangle + \frac{1}{\sqrt{2}}|\leftarrow\rangle. \tag{5.27}$$

Substituting this expression in (5.26) gives:

$$p\left(\frac{\hbar}{2}\right) = \left(\frac{1}{\sqrt{2}}\langle\rightarrow| + \frac{1}{\sqrt{2}}\langle\leftarrow|\right)(|\rightarrow\rangle\langle\rightarrow|)$$

$$\times \left(\frac{1}{\sqrt{2}}|\rightarrow\rangle + \frac{1}{\sqrt{2}}|\leftarrow\rangle\right) = \frac{1}{2}. \tag{5.28}$$

Similarly, the probability that the outcome of the measurement is $-\frac{\hbar}{2}$ is given by:

$$p\left(-\frac{\hbar}{2}\right) = \langle\uparrow|\,(|\leftarrow\rangle\langle\leftarrow|)\,|\uparrow\rangle \tag{5.29}$$

Substituting the change of basis expression (5.27) into this expression gives:

$$p\left(-\frac{\hbar}{2}\right) = \left(\frac{1}{\sqrt{2}}\langle\rightarrow| + \frac{1}{\sqrt{2}}\langle\leftarrow|\right)(|\leftarrow\rangle\langle\leftarrow|)$$

$$\times \left(\frac{1}{\sqrt{2}}|\rightarrow\rangle + \frac{1}{\sqrt{2}}|\leftarrow\rangle\right) = \frac{1}{2}. \tag{5.30}$$

Hence, both outcomes of the measurement of $J_1$ occur with equal probability, and the beam is split into two again.

> **Experiment 4:**

We consider a beam of spin $\frac{1}{2}$ particles in the initial state:

$$\psi = \alpha|\uparrow\rangle + \beta|\downarrow\rangle, \quad |\alpha|^2 + |\beta|^2 = 1. \tag{5.31}$$

The beam passes through a Stern-Gerlach apparatus where $J_3$ is measured, and particles where the outcome of the measurement is $\frac{\hbar}{2}$ are allowed to pass through a second Stern-Gerlach device where the magnetic field gradient is in the $x$ direction. Immediately before passing through the second Stern-Gerlach device the particles are in the state $|\uparrow\rangle$, and we wish to measure $J_1$. We saw from experiment 3 that these particles will be in state $|\rightarrow\rangle$ with probability $\frac{1}{2}$ and in state $|\leftarrow\rangle$ with probability $\frac{1}{2}$. The particles in the state $|\rightarrow\rangle$ (corresponding to a measurement outcome of $\frac{\hbar}{2}$) are allowed to continue to a third Stern-Gerlach apparatus where the gradient of the magnetic field is in the $z$ direction. We measure $J_3$ when the particles have passed through this final Stern-Gerlach apparatus. The possible outcomes are the eigenvalues of $J_3$, $\pm\frac{\hbar}{2}$. The probability that the outcome of the measurement is $\frac{\hbar}{2}$ is given by:

$$p\left(\frac{\hbar}{2}\right) = \langle\rightarrow|\,(|\uparrow\rangle\langle\uparrow|)\,|\rightarrow\rangle \tag{5.32}$$

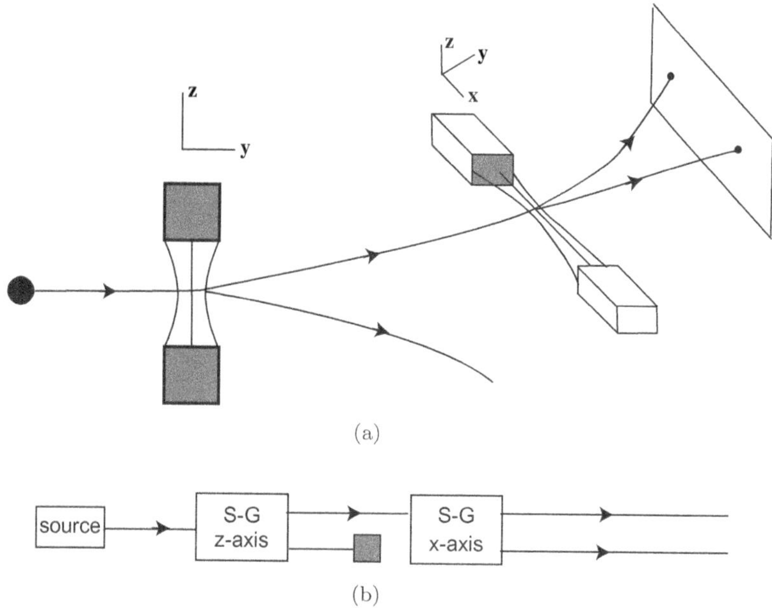


Figure 5.3: A beam of spin $\frac{1}{2}$ particles pass through two consecutive Stern-Gerlach devices. a) The first Stern-Gerlach device measures $J_3$, and after passing through the first Stern-Gerlach device only particles in the state $|\uparrow\rangle$ are allowed to pass through the second Stern-Gerlach device. The second Stern-Gerlach device measures $J_1$. b) A more "schematic" depiction of the experiment that is shown in a). Fig. 5.3b (from Wikipedia).

In order to calculate this quantity we need to express the state $|\rightarrow\rangle$ in terms of the basis of $\mathbb{C}^2$ given by the eigenvectors of $J_3$. Recall that this was given in (5.18), which we rewrite here:

$$|\rightarrow\rangle = \frac{1}{\sqrt{2}}|\uparrow\rangle + \frac{1}{\sqrt{2}}|\downarrow\rangle \tag{5.33}$$

Then we have:

$$p\left(\frac{\hbar}{2}\right) = \left(\frac{1}{\sqrt{2}}\langle\uparrow| + \frac{1}{\sqrt{2}}\langle\downarrow|\right)(|\uparrow\rangle\langle\uparrow|)\left(\frac{1}{\sqrt{2}}|\uparrow\rangle + \frac{1}{\sqrt{2}}|\downarrow\rangle\right) = \frac{1}{2}. \tag{5.34}$$

The probability that the outcome of the measurement is $-\frac{\hbar}{2}$ is given by:

$$p\left(-\frac{\hbar}{2}\right) = \langle\rightarrow|\,(|\downarrow\rangle\langle\downarrow|)\,|\rightarrow\rangle \tag{5.35}$$

A similar calculation to that given above shows that:

$$p\left(-\frac{\hbar}{2}\right) = \frac{1}{2}.$$

Hence, the final Stern-Gerlach devices splits the beam into two, corresponding to "spin up" and "spin down", as measured in the vertical ($z$) direction. This is exactly the situation that occurred immediately after passing through the first Stern-Gerlach device, where we "threw away" the "spin down" part of the beam. Somehow, some "spin down" component of the beam has now been recovered.

## 5.3 Composite Systems, Tensor Products, and Entanglement

*Motivation for considering the tensor product.* To begin with, we consider classical mechanics. Imagine a point particle of mass $m$ undergoing two dimensional motion as a result of a (net) force acting on the particle. In classical mechanics the phrase *configuration space* is used to describe the two coordinates that describe the position of the particle, which is two dimensional. Now consider $n$ separate and identical (without loss of generality for the point we now wish to make) such systems that we view as a single *composite system*. The configuration space for composite systems is the *cartesian product* of the configuration spaces of each two dimensional system, and it is $2n$ dimensional. As we have described it thus far, the individual two dimensional systems do not interact with each other. However, we could consider them to be coupled in a variety of ways. The nature of the coupling (i.e. the coupling forces) can be determined by applying Newton's law's to determine the forces due to the coupling of the individual systems. These coupling forces can be expressed in terms of the configuration space coordinates of the cartesian product of the individual systems. So even if the individual two dimensional systems are coupled, the configuration space for the $n$ coupled systems is the cartesian product of the configuration spaces of the individual two dimensional systems, and it is $2n$ dimensional.

Now let's consider the quantum mechanical analog of this situation. We consider $n$ quantum mechanical systems, each described by a two dimensional state space (Hilbert space). Now we consider the *composite system* of all $n$ two dimensional systems. In quantum mechanics the state space describing the composite system is *not* the cartesian product of the two dimensional state spaces (which would have dimension $2n$), but the

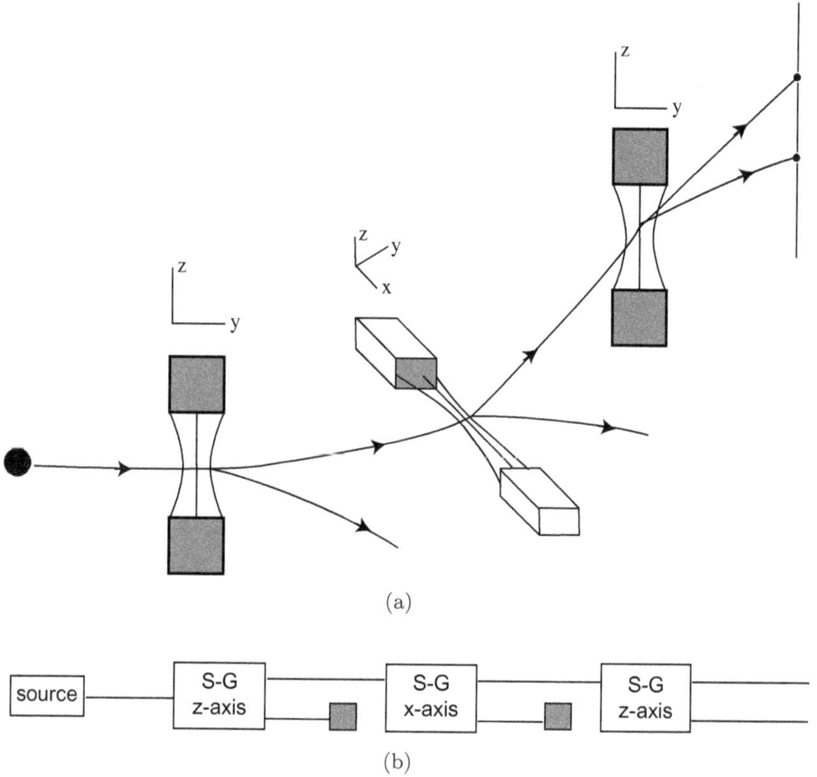

(a)

source — S-G z-axis — ▪ — S-G x-axis — ▪ — S-G z-axis —

(b)

Figure 5.4: A beam of spin $\frac{1}{2}$ particles pass through three consecutive Stern-Gerlach devices. a) The first Stern-Gerlach device measures $J_3$, and after passing through the first Stern-Gerlach device only particles in the state $|\uparrow\rangle$ are allowed to pass through the second Stern-Gerlach device. A measurement of $J_1$ is made on these particles, and only particles with outcome $\frac{\hbar}{2}$ are allowed to continue to the third Stern-Gerlach device. The third Stern-Gerlach device measures $J_3$. b) A more "schematic" depiction of the experiment that is shown in a). Fig 5.4b (from Wikipedia).

*tensor product* of the two dimensional state spaces, which has dimension $2^n$. This is a somewhat bewildering difference between classical and quantum mechanics and arises as a result of our consideration of composite (or coupled, or interacting) systems. The first point to take notice of is the difference in dimension for the composite system of $n$ two dimensional classical mechanical systems versus the composite system of $n$ two dimensional quantum mechanical systems. In classical mechanics the dimensions is $2n$ and in quantum mechanics the dimension is $2^n$ (more details of this will be provided when we consider tensor product spaces in detail below). Hence, as the number of two dimensional systems is increased, the dimension of the state space of the composite system grows linearly with $n$ in classical mechanics, and exponentially with $n$ in quantum mechanics. This is a huge difference between classical and quantum mechanics (and it is at the heart of why a quantum computer (assuming such a thing existed) could solve much more complex problems than a classical computer).

However, the real question is "why the difference?", i.e. why a cartesian product in classical mechanics and a tensor product in quantum mechanics? In classical mechanics Newton's laws are our guide to writing down the "equations of motion" for a classical mechanical system. Consider the situation of $n$ systems each exhibiting two-dimensional motion as described above. The solution of Newton's equations for the composite system is a time varying vector in the cartesian product of the state spaces of the individual two dimensional systems. In quantum mechanics the Schrödinger equation is our guide to writing down the equations of motion for a quantum mechanical system, and its solution is the wavefunction. For the composite system of $n$ two dimensional systems the solution of Schrödinger's equation is a wave function, which is a function of the $2n$ configuration space variables for the composite system. Mathematically, this wavefunction can be represented as the tensor product of wavefunctions for the individual two dimensional systems. Hence, the Schrödinger equation description of "quantum reality" implies the tensor product description of composite systems. A more detailed discussion of this argument is given in Notes 17 of the "Littlejohn Notes". However, you will not need to worry about this question of "why" in what follows (unless you want to). We will begin our study of composite systems by considering Hilbert spaces that are tensor products and linear operators defined on Hilbert spaces that are tensor products.

### *Tensor Product State Spaces and Operators on Tensor Product State Spaces*

We begin by defining the tensor product of two Hilbert spaces.[5] Finite dimensional Hilbert spaces will suffice for our needs (and the main example that we will use repeatedly is $\mathbb{C}^2$).

---

[5]Tensor products of a finite number of Hilbert spaces are straightforwardly defined, but they will not be required for our purposes

**Definition 30 (Tensor Product).** Let $\mathcal{H}_A$ and $\mathcal{H}_B$ denote finite dimensional, complex linear vector spaces, each equipped with an inner product (i.e. finite dimensional Hilbert spaces)[6]. The tensor product of $\mathcal{H}_A$ and $\mathcal{H}_B$, denoted $\mathcal{H}_A \otimes \mathcal{H}_B$ is a finite dimensional, complex linear vector space, equipped with an inner product, consisting of i) all elements of the form $|a\rangle \otimes |b\rangle$, where $|a\rangle \in \mathcal{H}_A$ and $|b\rangle \in \mathcal{H}_B$ and ii) all possible (finite) linear combinations of elements $|a\rangle \otimes |b\rangle$, where $|a\rangle \in \mathcal{H}_A$ and $|b\rangle \in \mathcal{H}_B$. The operations on elements in $\mathcal{H}_A \otimes \mathcal{H}_B$ obey the following rules.

1. $\lambda \left( |a\rangle \otimes |b\rangle \right) = (\lambda|a\rangle) \otimes |b\rangle = |a\rangle \otimes (\lambda|b\rangle), \quad \lambda \in \mathbb{C},$

2. $(|a_1\rangle + |a_2\rangle) \otimes |b\rangle = |a_1\rangle \otimes |b\rangle + |a_2\rangle \otimes |b\rangle,$

   $|a\rangle \otimes (|b_1\rangle + |b_2\rangle) = |a\rangle \otimes |b_1\rangle + |a\rangle \otimes |b_2\rangle$

3. Let $\langle \cdot | \cdot \rangle_A$ denote the inner product on $\mathcal{H}_A$, $\langle \cdot | \cdot \rangle_B$ denote the inner product on $\mathcal{H}_B$, and $\langle \cdot | \cdot \rangle_{A \otimes B}$ denote the inner product on $\mathcal{H}_A \otimes \mathcal{H}_B$. Then for $|a\rangle \otimes |b\rangle,\, |a'\rangle \otimes |b'\rangle \in \mathcal{H}_A \otimes \mathcal{H}_B$, the inner product of these two vectors is given by:

$$\left\langle \left( |a\rangle \otimes |b\rangle \right) \,\middle|\, \left( |a'\rangle \otimes |b'\rangle \right) \right\rangle_{A \otimes B} \equiv \langle a|a'\rangle_A \langle b|b'\rangle_B$$

There are some "mathematical holes" in our definition. We are *stating* that $\mathcal{H}_A \otimes \mathcal{H}_B$ is a finite dimensional, complex linear vector space. That statement requires a proof, which we will not give here. However, recall that a vector space must have a set of scalars associated with it (for us, these are the complex numbers) and a way of adding the objects that are elements of the vector space. Then addition of the objects and multiplication of the objects by scalars must satisfy certain axioms. It would be a useful exercise for you to prove that $\mathcal{H}_A \otimes \mathcal{H}_B$ satisfies the requirements to be a vector space.[7] Moreover, an inner product on a vector space must satisfy certain requirements, and it would be a useful exercise for you to show that the inner product defined above satisfies these requirements.

---

[6] It is worth commenting on the nature of the subscripts "A" and "B'. We will be discussing measurement, which requires a state space and an observable (i.e. Hermitian operator acting on that state space). Practically, measurements are carried out by people, and it is traditional to speak of "Alice" (A) and "Bob" (B) as two people carrying out measurements on a system. More people could be involved (e.g. C for "Charlie"), but we will not consider the situation of more than two people carrying out measurements.

[7] In order to show the existence of an additive identity element it would be useful to first show that $|a\rangle \in \mathcal{H}_A, |b\rangle \in \mathcal{H}_B$, we have $|\text{zero}\rangle_{\mathcal{H}_A} \otimes |b\rangle = |a\rangle \otimes |\text{zero}\rangle_{\mathcal{H}_B} = |\text{zero}\rangle_{\mathcal{H}_A \otimes \mathcal{H}_B}$, where $|\text{zero}\rangle_{\mathcal{H}_A}$ denotes the zero vector in $\mathcal{H}_A$, etc.

Next, we turn to the important issue of a basis for the tensor product space.

**Proposition 5 (Orthonormal Basis for the Tensor Product Space).** Let $\{|a_i\rangle\}$, $i = 1, \ldots, N_A$ be an orthonormal basis in $\mathcal{H}_A$ and let $\{|b_i\rangle\}$, $i = 1, \ldots, N_B$ be an orthonormal basis in $\mathcal{H}_B$. Then $|a_i\rangle \otimes |b_j\rangle$, $i = 1, \ldots, N_A$, $j = 1, \ldots N_B$ is an orthonormal basis in $\mathcal{H}_A \otimes \mathcal{H}_B$.

Note that it follows immediately from this result that the dimension of $\mathcal{H}_A \otimes \mathcal{H}_B$ is $N_A N_B$. The proof of this result is left as an exercise (and it is one that you should really make an effort to do).

**Example $\mathbb{C}^2 \otimes \mathbb{C}^2$.** As an example we construct a basis for $\mathbb{C}^2 \otimes \mathbb{C}^2$. We first choose a basis for $\mathbb{C}^2$:

$$\begin{pmatrix} 1 \\ 0 \end{pmatrix} \equiv |0\rangle, \qquad \begin{pmatrix} 0 \\ 1 \end{pmatrix} \equiv |1\rangle. \tag{5.36}$$

Recall from (5.2), (5.3) and (5.4) that these are the eigenvectors of the matrix $Z$. Following Proposition 5, an orthonormal basis for $\mathbb{C}^2 \otimes \mathbb{C}^2$ consists of the following four vectors (with a shorthand notation for each vector immediately to the right):

$$\begin{aligned}
|0\rangle \otimes |0\rangle &\equiv |0\rangle|0\rangle, \\
|0\rangle \otimes |1\rangle &\equiv |0\rangle|1\rangle, \\
|1\rangle \otimes |0\rangle &\equiv |1\rangle|0\rangle, \\
|1\rangle \otimes |1\rangle &\equiv |1\rangle|1\rangle.
\end{aligned} \tag{5.37}$$

Now we come to a very important idea. Let $|\psi\rangle \in \mathcal{H}_A \otimes \mathcal{H}_B$. It follows from Proposition 5 that $|\psi\rangle$ can be written as a linear combination of elements of $\mathcal{H}_A \otimes \mathcal{H}_B$ of the form $|a_i\rangle \otimes |b_j\rangle$. However, it is *not* true that *for every* $|\psi\rangle \in \mathcal{H}_A \otimes \mathcal{H}_B$ we can find $|\psi_A\rangle \in \mathcal{H}_A$, $|\psi_B\rangle \in \mathcal{H}_B$ such that:

$$|\psi\rangle = |\psi_A\rangle \otimes |\psi_B\rangle. \tag{5.38}$$

We will show that this is the case by considering a specific example in $\mathbb{C}^2 \otimes \mathbb{C}^2$ that *cannot* be represented in product form.

Consider the state:

$$|\phi\rangle = \frac{|0\rangle|0\rangle + |1\rangle|1\rangle}{\sqrt{2}} \tag{5.39}$$

We wish to find states $|\phi_A\rangle$ and $|\phi_B\rangle$ such that $|\phi\rangle = |\phi_A\rangle \otimes |\phi_B\rangle$. Letting

$$|\phi_A\rangle = a_0|0\rangle + a_1|1\rangle, \qquad |\phi_B\rangle = b_0|0\rangle + b_1|1\rangle$$

Then we have:

$$|\phi_A\rangle \otimes |\phi_B\rangle = (a_0|0\rangle + a_1|1\rangle) \otimes (b_0|0\rangle + b_1|1\rangle)$$

$$= a_0 b_0|0\rangle|0\rangle + a_0 b_1|0\rangle|1\rangle + a_1 b_0|1\rangle|0\rangle + a_1 b_1|1\rangle|1\rangle. \qquad (5.40)$$

Then if $(5.39) = (5.40)$ we must have:

$$\frac{1}{\sqrt{2}} = a_0 b_0$$

$$0 = a_0 b_1$$

$$0 = a_1 b_0$$

$$\frac{1}{\sqrt{2}} = a_1 b_1. \qquad (5.41)$$

If we examine the second equality in this list, $0 = a_0 b_1$ implies either $a_0 = 0$ or $b_1 = 0$. However, either of these conditions being satisfied is inconsistent with the first and the fourth inequalities in this list. So we cannot find states $|\phi_A\rangle$ and $|\phi_B\rangle$ such that $|\phi\rangle = |\phi_A\rangle \otimes |\phi_B\rangle$. This leads to the following definition.

**Definition 31 (Entangled State).** Any state $|\phi\rangle \in \mathcal{H}_A \otimes \mathcal{H}_B$ that cannot be written in the form $|\phi\rangle = |\phi_A\rangle \otimes |\phi_B\rangle$, for some $|\phi_A\rangle \in \mathcal{H}_A$, $|\phi_B\rangle \in \mathcal{H}_B$ is said to be an *entangled state*.

The following definition should be clear.

**Definition 32 (Product State).** Any state $|\phi\rangle \in \mathcal{H}_A \otimes \mathcal{H}_B$ that can be written in the form $|\phi\rangle = |\phi_A\rangle \otimes |\phi_B\rangle$, for some $|\phi_A\rangle \in \mathcal{H}_A$, $|\phi_B\rangle \in \mathcal{H}_B$ is said to be a *product state*.

**Linear Operators on Tensor Product Spaces.** Suppose $A : \mathcal{H}_A \to \mathcal{H}_A$ and $B : \mathcal{H}_B \to \mathcal{H}_B$ are linear operators. We construct a linear operator on $\mathcal{H}_A \otimes \mathcal{H}_B$ using $A$ and $B$, which we will refer to as $A \otimes B$, as follows.

Recall Definition 5. It suffices to define a linear operator of a complex linear vector space by specifying its action of each basis element. Then the action of the linear map on a general vector follows by linearity since any vector can be expressed as a linear combination of basis vectors. We define:

$$(A \otimes B)|a_i\rangle|b_j\rangle \equiv A|a_i\rangle B|b_j\rangle, \quad i = 1, \ldots N_A, \; j = 1, \ldots N_B. \qquad (5.42)$$

Now choose a general $|\psi\rangle \in \mathcal{H}_A \otimes \mathcal{H}_B$. Then we can write:

$$|\psi\rangle = \sum_{i,j} \alpha_{i,j}|a_i\rangle|b_j\rangle.$$

Then we have:

$$(A \otimes B)|\psi\rangle = \sum_{i,j} \alpha_{i,j} A|a_i\rangle B|b_j\rangle. \tag{5.43}$$

The following result describes the eigenvalues and eigenvectors of $A \otimes B$ in terms of the eigenvalues and eigenvectors of $A$ and the eigenvalues and eigenvectors of $B$.

**Proposition 6.** Suppose the linear operator $A$ has eigenvalues $\lambda_i$ and eigenvectors $|a_i\rangle$ and suppose the linear operator $B$ has eigenvalues $\mu_i$ and eigenvectors $|b_i\rangle$. Then the linear operator $A \otimes B$ has eigenvalues $\lambda_i \mu_j$ and eigenvectors $|a_i\rangle|b_j\rangle$.

**Proof.** Verification of this result involves a straightforward calculation:

$$A \otimes B|a_i\rangle|b_j\rangle = A|a_i\rangle\, B|b_j\rangle,$$

$$= \lambda_i|a_i\rangle\, \mu_j|b_j\rangle,$$

$$= \lambda_i \mu_j |a_i\rangle|b_j\rangle. \tag{5.44}$$

$\square$

Now the natural question arises. If $A$ and $B$ have certain properties (e.g. if they are Hermitian), then does $A \otimes B$ have the same property? We have the following proposition.

**Proposition 7.**

$$(A \otimes B)^\dagger = A^\dagger \otimes B^\dagger.$$

**Proof.** The idea for the proof is that we show that this equality holds on basis vectors $|a_i\rangle|b_j\rangle$, and then it follows that it holds on an arbitrary vector since any vector can be expressed as a linear combination of basis vectors and the map is linear. We have:

$$\langle a_i|\langle b_j|A^\dagger \otimes B^\dagger |a_k\rangle|b_l\rangle = \langle a_i|A^\dagger|a_k\rangle\langle b_j|B^\dagger|b_l\rangle,$$

$$= (\langle a_k|A|a_i\rangle\langle b_l|B|b_j\rangle)^*,$$

$$= (\langle a_k|\langle b_l|A \otimes B|a_i\rangle|b_j\rangle)^*$$

$$= \langle a_i|\langle b_j|(A \otimes B)^\dagger |a_k\rangle|b_l\rangle \tag{5.45}$$

$\square$

This result implies that if $A$ and $B$ are Hermitian, then $A \otimes B$ is Hermitian. Using this result, together with Proposition 6 allows us to conclude that $A \otimes B$ has a spectral decomposition analogous to what we have derived earlier. In order to keep the discussion simple (and this is the most general case that we will need in this book), let us assume that $\mathcal{H}_A$ and $\mathcal{H}_B$ are finite dimensional, and $A$ and $B$ are nondegenerate. In this case, for $A$ there is a unique eigenvector, $|a_i\rangle$, for every eigenvalue $\lambda_i$, $i = 1, \ldots N_A$, and for $B$ there is a unique eigenvector, $|b_i\rangle$, for every eigenvalue $\mu_i$, $i = 1, \ldots N_B$. Then we have:

$$A \otimes B = \sum_{i,j} \lambda_i \mu_j |a_i\rangle |b_j\rangle \langle a_i| \langle b_j|,$$

$$= \left( \sum_i \lambda_i |a_i\rangle \langle a_i| \right) \otimes \left( \sum_j \mu_j |b_j\rangle \langle b_j| \right). \tag{5.46}$$

The following result will be useful for computations.

**Proposition 8.**

$$(A \otimes B)(C \otimes D) = AC \otimes BD.$$

**Proof.** The idea is the same as the idea for the proof of Proposition 7. We show that this equality holds on basis vectors $|a_i\rangle |b_j\rangle$, and then it follows that it holds on an arbitrary vector since any vector can be expressed as a linear combination of basis vectors and the map is linear. We have:

$$(A \otimes B)(C \otimes D) |a_i\rangle |b_j\rangle = (A \otimes B)(C|a_i\rangle \otimes D|b_j\rangle),$$

$$= AC|a_i\rangle \, BD|b_j\rangle,$$

$$= (AC \otimes BD) |a_i\rangle |b_j\rangle. \tag{5.47}$$

$\square$

**Example: The Pauli Spin Operators and Tensor Products** Recall the matrices $X$ and $Z$ associated with the Pauli spin matrices defined on $\mathbb{C}^2$ given in (5.2). We denote the basis of $\mathbb{C}^2$ by $|0\rangle$ and $|1\rangle$ (recall (5.3) and (5.4)). Then we have:

$$X|0\rangle = |1\rangle, \; Z|0\rangle = |0\rangle,$$
$$X|1\rangle = |0\rangle, \; Z|1\rangle = -|1\rangle. \tag{5.48}$$

We define the tensor product operator $X \otimes Z$ on $\mathbb{C}^2 \otimes \mathbb{C}^2$ by defining the action of the operator on the basis elements of $\mathbb{C}^2 \otimes \mathbb{C}^2$ as follows:

$$X \otimes Z|0\rangle|0\rangle = X|0\rangle Z|0\rangle = |1\rangle|0\rangle,$$
$$X \otimes Z|1\rangle|1\rangle = X|1\rangle Z|1\rangle = -|0\rangle|1\rangle,$$
$$X \otimes Z|0\rangle|1\rangle = X|0\rangle Z|1\rangle - -|1\rangle|1\rangle,$$
$$X \otimes Z|1\rangle|0\rangle = X|1\rangle Z|0\rangle = |0\rangle|0\rangle.$$

Then an example of the action of $X \otimes Z$ on a general state in $\mathbb{C}^2 \otimes \mathbb{C}^2$ is the following:

$$X \otimes Z \left( \frac{|0\rangle|0\rangle + |1\rangle|1\rangle}{\sqrt{2}} \right) = \frac{|1\rangle|0\rangle - |0\rangle|1\rangle}{\sqrt{2}}. \tag{5.49}$$

At this point it is useful to note that $X \otimes Z \neq Z \otimes X$. In order to show this it suffices to show that inequality holds on only one element of $\mathbb{C}^2 \otimes \mathbb{C}^2$:

$$X \otimes Z|0\rangle|0\rangle = |1\rangle|0\rangle,$$

and

$$Z \otimes X|0\rangle|0\rangle = |0\rangle|1\rangle.$$

**Example.** The operator corresponding to doing nothing is the identity operator:

$$\mathbb{I}|\psi\rangle = |\psi\rangle, \quad \forall|\psi\rangle. \tag{5.50}$$

Then $X \otimes \mathbb{I}$ is a tensor product operator on $\mathbb{C}^2 \otimes \mathbb{C}^2$. Examples of the action of this operator on states in $\mathbb{C}^2 \otimes \mathbb{C}^2$ are the following:

$$X \otimes \mathbb{I}|0\rangle|0\rangle = |1\rangle|0\rangle. \tag{5.51}$$

$$X \otimes \mathbb{I} \left( \frac{|0\rangle|0\rangle + |1\rangle|1\rangle}{\sqrt{2}} \right) = \frac{|1\rangle|0\rangle + |0\rangle|1\rangle}{\sqrt{2}}. \tag{5.52}$$

## 5.4 Measurement: Observables as Tensor Product Operators on Tensor Product Spaces.

We now consider the issue of measurement, where the observable is the tensor product of two Hermitian operators on a Hilbert space that is the tensor product of the two Hilbert spaces on which the individual operators making up the tensor product of the two operators acts (you might want to think about this sentence a bit).

However, let's first consider $Z$ acting on $\mathbb{C}^2$. We wish to measure $Z$ in the state

$$|\psi\rangle = \frac{|0\rangle + |1\rangle}{\sqrt{2}} \in \mathbb{C}^2.$$

Recall from (5.7) that the spectral decomposition of $Z$ is given by:

$$Z = |0\rangle\langle 0| - |1\rangle\langle 1|,$$

$P_0 \equiv |0\rangle\langle 0|$ is the projection operator onto the eigenstate corresponding to the eigenvalue $+1$ and $P_1 \equiv |1\rangle\langle 1|$ is the projection operator onto the eigenstate corresponding to the eigenvalue $-1$. The post-measurement state given that the outcome of the measurement is $+1$ is given by:

$$\frac{P_0|\psi\rangle}{\|P_0|\psi\rangle\|} = \frac{\frac{1}{\sqrt{2}}|0\rangle\langle 0|\left(|0\rangle + |1\rangle\right)}{\|\frac{1}{\sqrt{2}}|0\rangle\langle 0|\left(|0\rangle + |1\rangle\right)\|},$$

$$= \frac{\frac{1}{\sqrt{2}}|0\rangle}{\|\frac{1}{\sqrt{2}}|0\rangle\|},$$

$$= |0\rangle, \tag{5.53}$$

and the probability that the outcome of the measurement is $+1$ is given by:

$$\langle\psi|P_0|\psi\rangle = \frac{1}{2}\left(\langle 0| + \langle 1|\right)\left(|0\rangle\langle 0|\right)\left(|0\rangle + |1\rangle\right),$$

$$= \frac{1}{2}\left(\langle 0| + \langle 1|\right)\left(|0\rangle\right) = \frac{1}{2}. \tag{5.54}$$

Similarly, the post-measurement state, given that the outcome of the measurement is $-1$, is given by:

$$\frac{P_1|\psi\rangle}{\|P_1|\psi\rangle\|} = |1\rangle, \tag{5.55}$$

and the probability that the outcome of the measurement is $-1$ is given by:

$$\langle\psi|P_1|\psi\rangle = \frac{1}{2}. \tag{5.56}$$

Now let's consider the operator $Z\otimes\mathbb{I}$ on $\mathbb{C}^2\otimes\mathbb{C}^2$, and we wish to measure $Z\otimes\mathbb{I}$ in the state,

$$|\psi\rangle = \frac{|0\rangle|0\rangle + |1\rangle|1\rangle}{\sqrt{2}} \in \mathbb{C}^2\otimes\mathbb{C}^2. \tag{5.57}$$

We know that $Z$ has eigenvalues $+1$ and $-1$, and $\mathbb{I}$ has eigenvalues $+1$ and $+1$. Therefore by Proposition 6 $Z\otimes\mathbb{I}$ has eigenvalues $+1$, $+1$, $-1$, and $-1$. Moreover, from (5.46) we have the following spectral representation:

$$Z\otimes\mathbb{I} = \left(|0\rangle\langle 0| - |1\rangle\langle 1|\right)\otimes\mathbb{I}$$

$$= |0\rangle\langle 0|\otimes\mathbb{I} - |1\rangle\langle 1|\otimes\mathbb{I}, \tag{5.58}$$

where $P_0 \equiv |0\rangle\langle 0|\otimes\mathbb{I}$ is the projection onto the space of eigenvectors corresponding to the eigenvalue $+1$ and $P_1 \equiv |1\rangle\langle 1|\otimes\mathbb{I}$ is the projection onto the space of eigenvectors corresponding to the eigenvalue $-1$.

Therefore, the post-measurement state given that the outcome of the measurement is $+1$ is given by:

$$\frac{P_0|\psi\rangle}{\|P_0|\psi\rangle\|} = \frac{(|0\rangle\langle 0| \otimes \mathbb{I})\frac{1}{\sqrt{2}}(|0\rangle|0\rangle + |1\rangle|1\rangle)}{\|P_0|\psi\rangle\|},$$

$$= \frac{\frac{1}{\sqrt{2}}|0\rangle|0\rangle}{\|\frac{1}{\sqrt{2}}|0\rangle|0\rangle\|} = |0\rangle|0\rangle, \tag{5.59}$$

and the probability that the outcome of the measurement is $+1$ is given by:

$$\langle\psi|P_0|\psi\rangle = \langle\psi|(|0\rangle\langle 0| \otimes \mathbb{I})\frac{1}{\sqrt{2}}(|0\rangle|0\rangle + |1\rangle|1\rangle),$$

$$= \frac{1}{2}((\langle 0|\langle 0| + \langle 1|\langle 1|)|0\rangle|0\rangle = \frac{1}{2}, \tag{5.60}$$

Similar calculations can be carried out for the post-measurement state given that the outcome of the measurement is $-1$ and the probability that the outcome of the measurement is $-1$.

## 5.5 Non-Locality and Bell Inequalities

### *The Einstein-Podolsky-Rosen (EPR) Paradox*

The thought experiment of Einstein, Podolsky, and Rosen has played a fundamental role in our understanding of quantum mechanics. The original reference is the following:

A. Einstein, B. Podolsky, and N. Rosen, Can quantum-mechanical description of physical reality be considered complete? *Phys. Rev.*, **47**, 777 (1935).

We will first describe the thought experiment, and discuss what were believed to be "paradoxical features" of the outcome of the thought experiments afterwards.

Suppose we have a source that creates two quantum particles of spin $\frac{1}{2}$. One is sent to Alice and one is sent to Bob ("A" and "B"). The two particles that are sent to Alice and Bob are created by the source through some type of physical process. Consequently, the two particles form a composite system. Figure 5.6 shows a schematic diagram of the two particles created by the source, with one sent to Alice and one sent to Bob. Alice and

Bob will each perform measurements on their particles. We assume that Alice and Bob are sufficiently spatially separated that no communication between them is possible in the time that it takes them to perform their measurements (even at the speed of light).

The Hilbert space for the composite system of two spin $\frac{1}{2}$ particles is $\mathbb{C}^2 \otimes \mathbb{C}^2$, and we assume that the two particles are in the following state (you might recall this state from earlier in the material on tensor products):

$$|\psi\rangle = \frac{1}{\sqrt{2}}(|0\rangle|0\rangle + |1\rangle|1\rangle). \tag{5.61}$$

Let us suppose that Alice measures $Z$, and Bob does not do anything. Then the measurement operator on $\mathbb{C}^2 \otimes \mathbb{C}^2$ is $Z \otimes \mathbb{I}$ (and we know that it is Hermitian). The spectral decomposition of this operator is given by:

$$Z \otimes \mathbb{I} = \underbrace{|0\rangle\langle 0| \otimes \mathbb{I}}_{P_0} - \underbrace{|1\rangle\langle 1| \otimes \mathbb{I}}_{P_1},$$

where $P_0$ is the projection onto the eigenspace corresponding to eigenvalues of $+1$ and $P_1$ is the projection onto the eigenspace corresponding to eigenvalues of $-1$.

Using the postulate of quantum mechanics concerned with measurement, if the outcome of Alice's measurement is $+1$ then the state collapses to:

$$\frac{P_0|\psi\rangle}{\|P_0|\psi\rangle\|} = \frac{(|0\rangle\langle 0| \otimes \mathbb{I})\left(\frac{1}{\sqrt{2}}(|0\rangle|0\rangle + |1\rangle|1\rangle)\right)}{\|(|0\rangle\langle 0| \otimes \mathbb{I})\left(\frac{1}{\sqrt{2}}(|0\rangle|0\rangle + |1\rangle|1\rangle)\right)\|} = |0\rangle|0\rangle.$$

$$\tag{5.62}$$

If the outcome of Alice's measurement is $-1$ then the state collapses to:

$$\frac{P_1|\psi\rangle}{\|P_1|\psi\rangle\|} = \frac{(|1\rangle\langle 1| \otimes \mathbb{I})\left(\frac{1}{\sqrt{2}}(|0\rangle|0\rangle + |1\rangle|1\rangle)\right)}{\|(|1\rangle\langle 1| \otimes \mathbb{I})\left(\frac{1}{\sqrt{2}}(|0\rangle|0\rangle + |1\rangle|1\rangle)\right)\|} = |1\rangle|1\rangle.$$

$$\tag{5.63}$$

It is important to realize that Alice's measurement has collapsed the wave function *for the composite system.*

Figure 5.5: Schematic of the source creating two particles, one is sent to Alice and one is sent to Bob.

After Alice has completed her measurement (and therefore collapsed the wave function) suppose that Bob measures $Z$. In this case the measurement operator for the composite system is given by:

$$\mathbb{I} \otimes Z = \underbrace{\mathbb{I} \otimes |0\rangle\langle 0|}_{P_0'} - \underbrace{\mathbb{I} \otimes |1\rangle\langle 1|}_{P_1'}, \tag{5.64}$$

where $P_0'$ is the projection onto the eigenspace corresponding to $+1$ and $P_1'$ is the projection onto the eigenspace corresponding to $-1$.

Now if Alice obtained $+1$ for her measurement of $Z$ then the state collapses to $|0\rangle|0\rangle$ and the probabilities that Bob measures $1$ or $-1$ are given by:

$$\begin{aligned}
p\,(\text{Bob obtains} + 1) &= \langle 0|\langle 0|P_0'|0\rangle|0\rangle = 1, \\
p\,(\text{Bob obtains} - 1) &= \langle 0|\langle 0|P_1'|0\rangle|0\rangle = 0.
\end{aligned} \tag{5.65}$$

If Alice obtained $-1$ for her measurement of $Z$ then the state collapses to $|1\rangle|1\rangle$ and the probabilities that Bob measures $1$ or $-1$ are given by:

$$\begin{aligned}
p\,(\text{Bob obtains} + 1) &= \langle 1|\langle 1|P_0'|1\rangle|1\rangle = 0, \\
p\,(\text{Bob obtains} - 1) &= \langle 1|\langle 1|P_1'|1\rangle|1\rangle = 1.
\end{aligned} \tag{5.66}$$

So Bob always obtains the same result as Alice.

One could ask if this would still hold if Alice measured $X$, and then Bob measured $X$? In this case the measurement operators are given by:

$$X \otimes \mathbb{I} = |+\rangle\langle +| \otimes \mathbb{I} - |-\rangle\langle -| \otimes \mathbb{I}, \quad \text{Alice},$$

$$\mathbb{I} \otimes X = \mathbb{I} \otimes |+\rangle\langle +| - \mathbb{I} \otimes |-\rangle\langle -| \otimes \mathbb{I}, \quad \text{Bob}. \tag{5.67}$$

We need to express $|\psi\rangle$ in terms of the basis of eigenvectors of $X$. Recall from (5.17) we have:

$$|0\rangle = \frac{|+\rangle + |-\rangle}{\sqrt{2}}, \quad |1\rangle = \frac{|+\rangle - |-\rangle}{\sqrt{2}}, \tag{5.68}$$

and therefore

$$\begin{aligned}
|\psi\rangle &= \frac{1}{\sqrt{2}}\left(|0\rangle|0\rangle + |1\rangle|1\rangle\right), \\
&= \frac{1}{\sqrt{2}}\left(\frac{|+\rangle + |-\rangle}{\sqrt{2}}\right)\left(\frac{|+\rangle + |-\rangle}{\sqrt{2}}\right) \\
&\quad + \frac{1}{\sqrt{2}}\left(\frac{|+\rangle - |-\rangle}{\sqrt{2}}\right)\left(\frac{|+\rangle - |-\rangle}{\sqrt{2}}\right), \\
&= \frac{1}{\sqrt{2}}\left(|+\rangle|+\rangle + |-\rangle|-\rangle\right). 
\end{aligned} \tag{5.69}$$

It follows that we will obtain the same results for measuring $X$ as we did for $Z$ simply by realizing that interchanging the kets $|0\rangle \leftrightarrow |+\rangle$ and $|1\rangle \leftrightarrow |-\rangle$ gives same results. Therefore, if Alice first measures $X$ and Bob measures $X$ afterwards, he will obtain exactly the same results as Alice.

Note that $|\psi\rangle$ is entangled. Would the same conclusions hold if the two particles were not entangled?

In summary the postulate of measurement in quantum mechanics says that if Alice measures a quantity then the state collapses to the state corresponding to the eigenstate of the outcome of Alice's measurement. Therefore if Bob measures the same quantity he will obtain exactly the same value as Alice. The result of this experiment seems contrary to common experience in two ways.

*Locality.* The experiment has been arranged to that no communication is possible between Alice and Bob, after Alice makes her measurement, and before Bob makes his measurement. ("Faster than the speed of light" communications violate special relativity.) Yet, somehow, at the instant Alice makes her measurement Bob's wave function "collapses" to the eigenstate corresponding to the eigenvalue that Alice measured.

*Reality.* We believe that the properties of a particle should be an intrinsic characteristic of the particle and they should *not* depend on what measurements are made. Once Alice makes her measurement of $Z$, if Bob then measures $Z$ he will always obtain the same value as Alice.

These issues troubled Einstein greatly. In letters to Niels Bohr he coined the famous phrase "spooky action at a distance"[8] to describe the results of this thought experiment. Einstein sought a way around these issues, and that led to "local hidden variable models", that we now describe.

### Local Hidden Variable Models

We give a brief description of the idea behind "hidden variable theories". Additional details can be found in Sakurai and Napolitano.[9] Suppose that in addition to the wave function, the particles had a "hidden" list of

---

[8] Max Born, Hedwig Born, Irene Born, and Albert Einstein. *The Born-Einstein letters: correspondence between Albert Einstein and Max and Hedwig Born from 1916 to 1955.* Walker, 1971.

[9] Jun John Sakurai and Jim Napolitano. *Modern quantum mechanics.* Cambridge University Press, 2020.

"answers" to measurements, and all the measurement does is reveal this answer.

Suppose that the EPR source produces pairs of particles that carry the same hidden variables. This would explain why Alice and Bob always get the same outcome, without any mysterious wave function collapse or instantaneous action at a distance, i.e. the particles carry hidden variables along with them, and the measurements are just revealing a "local" property of the particles. If the source sometimes produces pairs with one set of hidden variables, and sometimes another set (but always the same values for a given pair of particles, i.e. the two particles that were produced at the same time by the source), this would also explain why sometimes $+1$ is measured and sometimes $-1$ is measured.

Einstein believed that there were hidden variables which, if we knew their values, would allow us to predict the outcomes of their measurements exactly. Therefore "local realism" would be restored. Effectively, this means that quantum mechanics would be a statistical theory, much like classical statistical mechanics describes the behaviour of gasses probabilistically, even though the underlying classical mechanics is deterministic. However, Bohr believed that wavefunction collapse and probabilistic outcomes were a fundamental part of how nature works. Both of these explanations were consistent with the EPR thought experiment, and this is the way the situation remained for some years, until John Bell came up with a twist on the EPR experiment that allowed hidden variables models to be put to the test,[10] and this is the next topic that we will consider.

## 5.6 Bell's Experiment (The Clauser-Horne-Shimony-Holt (CHSH) Version)

Rather than discuss Bell's experiment, we will consider a slightly different, but similar in spirit, experiment due to Clauser, Horne, Shimony and Holt.[11] They described another thought experiment that, in principle, could be realized in the laboratory (we will mention this a bit more later).

---

[10] John S Bell. On the Einstein Podolsky Rosen paradox. *Physics Physique Fizika*, 1 (3):195, 1964.

[11] John F Clauser, Michael A Horne, Abner Shimony, and Richard A Holt. Proposed experiment to test local hidden-variable theories. *Physical review letters*, 23(15):880, 1969.

- A source creates a pair of particles. One is sent to Alice, and the other to Bob. Note that we have not specified whether or not these are classical particles or quantum particles.
- Alice has a choice of *two* properties, denoted $A_1$ and $A_2$, that she can measure on her particle, and the possible outcome of either measurement is $\pm 1$. More precisely, we denote the possible values of the property $A_1$ by $a_1 = \pm 1$ and the possible values of the property $A_2$ by $a_2 = \pm 1$. Similarly, Bob has a choice of measuring two properties, $B_1$ and $B_2$, that take possible values $b_1 = \pm 1$ and $b_2 = \pm 1$, respectively. We assume that what Alice decides to measure has no effect on what Bob decides to measure. This is the "locality assumption".
- Alice and Bob repeat the experiment many times (i.e. many particle pairs are sent to Alice and Bob from the source), and they choose at random which property that they will measure. They keep a record of both their measurement choices (i.e. the property they choose to measure) and the outcomes of the measurement.
- We denote the expected value of the product of $a_1 b_1$ $(= \pm 1)$ by $\mathbb{E}(a_1 b_1)$, the expected value of the product of $a_1 b_2$ $(= \pm 1)$ by $\mathbb{E}(a_1 b_2)$, the expected value of the product of $a_2 b_1$ $(= \pm 1)$ by $\mathbb{E}(a_2 b_1)$, and the expected value of the product of $a_1 b_1$ $(= \pm 1)$ by $\mathbb{E}(a_2 b_2)$. We assume that there is a joint probability distribution that governs the outcome of all measurements that Alice and Bob might perform, $P(a_1.a_2, b_1, b_2)$. This is the hypothesis of "reality". If the values of the outcomes are are known exactly, then the outcome of any measurement can be predicted with certainty–the measurement outcomes are described probabilistically because the values of the hidden variables are drawn from an ensemble of possible values.

Based on these assumptions, Clauser, Horne, Shimony and Holt (CHSH) derived an inequality that the experimental results must satisfy, under the assumptions above.

**Theorem 12 (The CHSH Inequality).**

$$\mathbb{E}(a_1 b_1) + \mathbb{E}(a_1 b_2) + \mathbb{E}(a_2 b_1) - \mathbb{E}(a_2 b_2) \leq 2$$

**Proof.** Consider the quantity:

$$C = a_1 b_1 + a_1 b_2 + a_2 b_1 - a_2 b_2,$$

$$= a_1(b_1 + b_2) + a_2(b_1 - b_2). \tag{5.70}$$

(Note that in writing (5.70) down we are assuming that values of all four observables can be specified simultaneously, and this is an assumption of

the "hidden variable theories" and not consistent with quantum mechanics.)
Since $b_1$, $b_2 = \pm 1$ we have

$$\text{either } b_1 + b_2 = \pm 2 \text{ and } b_1 - b_2 = 0,$$
$$\text{or } b_1 + b_2 = 0 \quad \text{and } b_1 - b_2 = \pm 2. \tag{5.71}$$

And also $a_1$, $a_2 = \pm 1$. Using these relations in (5.70) gives:

$$C = a_1(b_1 + b_2) + a_2(b_1 - b_2),$$
$$= (\pm 1)(\pm 2),$$
$$= \pm 2 \le 2. \tag{5.72}$$

Let $P(a_1, a_2, b_1, b_2)$ be the probability that the particles are in the state
where property $A_1$ has the value $a_1$, property $A_2$ has the value $a_2$, property
$B_1$ has the value $b_1$, and property $B_2$ has the value $b_2$.
Then the expected value for the quantity $C$ is given by:

$$\mathbb{E}(C) = \sum_{a_1, a_2, b_1, b_2} \left( a_1(b_1 + b_2) + a_2(b_1 - b_2) \right) P(a_1, a_2, b_1, b_2),$$

$$\le \sum_{a_1, a_2, b_1, b_2} 2\, P(a_1, a_2, b_1, b_2), \quad \text{using} \quad (5.72).$$

$$= 2. \tag{5.73}$$

But we also have:

$$\mathbb{E}(C) = \mathbb{E}\left( a_1 b_1 + a_1 b_2 + a_2 b_1 - a_2 b_2 \right)$$
$$= \mathbb{E}(a_1 b_1) + \mathbb{E}(a_1 b_2) + \mathbb{E}(a_2 b_1) - \mathbb{E}(a_2 b_2) \tag{5.74}$$

Combining this expression, with (5.73), gives:

$$\mathbb{E}(a_1 b_1) + \mathbb{E}(a_1 b_2) + \mathbb{E}(a_2 b_1) - \mathbb{E}(a_2 b_2) \le 2. \tag{5.75}$$

$$\square$$

Now let's consider this experiment with quantum particles. Suppose the
source emits pairs of quantum particles of spin $\frac{1}{2}$. For each pair emitted,
one is sent to Alice and one is sent to Bob. Moreover, we assume that each
emitted pair of particles is in the state:

$$|\psi\rangle = \frac{|0\rangle|0\rangle + |1\rangle|1\rangle}{\sqrt{2}} \tag{5.76}$$

(Note: this state should be familiar to you.)

Alice chooses to measure $A_1 = X$ or $A_2 = Z$ and Bob chooses to measure $B_1 = \frac{X+Z}{\sqrt{2}}$ or $B_2 = \frac{X-Z}{\sqrt{2}}$. Recall that:

$$X = \begin{pmatrix} 0 & 1 \\ 1 & 0 \end{pmatrix}, \quad Z = \begin{pmatrix} 1 & 0 \\ 0 & -1 \end{pmatrix},$$

and therefore

$$\frac{X+Z}{\sqrt{2}} = \frac{1}{\sqrt{2}} \begin{pmatrix} 1 & 1 \\ 1 & -1 \end{pmatrix}, \quad \frac{X-Z}{\sqrt{2}} = \frac{1}{\sqrt{2}} \begin{pmatrix} -1 & 1 \\ 1 & 1 \end{pmatrix}.$$

It is easy to verify that each of these four matrices has eigenvalues $\pm 1$. Therefore the outcome of any measurement of these four quantities is $\pm 1$.

We will need the following two calculations:

$$\langle \psi | Z \otimes Z | \psi \rangle = \frac{1}{2} \left( \langle 0 | \langle 0 | + \langle 1 | \langle 1 | \right) Z \otimes Z \left( | 0 \rangle | 0 \rangle + | 1 \rangle | 1 \rangle \right),$$

$$= \frac{1}{2}(1+1) = 1, \tag{5.77}$$

and

$$\langle \psi | X \otimes X | \psi \rangle = \frac{1}{2} \left( \langle 0 | \langle 0 | + \langle 1 | \langle 1 | \right) X \otimes X \left( | 0 \rangle | 0 \rangle + | 1 \rangle | 1 \rangle \right) = 1. \tag{5.78}$$

Now we have:

$$\mathbb{E}(a_1 b_1) + \mathbb{E}(a_1 b_2) + \mathbb{E}(a_2 b_1) - \mathbb{E}(a_2 b_2)$$

$$= \langle \psi | X \otimes \frac{X+Z}{\sqrt{2}} | \psi \rangle + \langle \psi | X \otimes \frac{X-Z}{\sqrt{2}} | \psi \rangle$$

$$+ \langle \psi | Z \otimes \frac{X+Z}{\sqrt{2}} | \psi \rangle - \langle \psi | Z \otimes \frac{X-Z}{\sqrt{2}} | \psi \rangle$$

$$= \frac{1}{\sqrt{2}} \left( 2 \langle \psi | X \otimes X | \psi \rangle + 2 \langle \psi | Z \otimes Z | \psi \rangle \right)$$

$$= \frac{4}{\sqrt{2}} = 2\sqrt{2} > 2 \tag{5.79}$$

But this result contradicts Theorem 12. Therefore there must have been assumptions in the proof of this inequality that are not valid for quantum mechanics. Recall that there were two key assumptions–the locality assumption and the "realism" assumption. One, or both, of these assumptions are not consistent with quantum mechanics. However, violation of the CHSH inequality rules out the hidden variable model as an explanation for the results of the EPR experiment.

Note that $|\psi\rangle$ is entangled. Would the CHSH inequality be violated if $|\psi\rangle$ were not entangled?

The experiments that we have described have been "thought experiments". However, technology has developed to the point where they can be realized in a laboratory. The main difficulties have been creating the pairs of entangled particles and ensuring that the locality assumption is enforced. There has been remarkable progress in this area, culminating in the Nobel Prize in Physics in 2022.[12] Wikipedia has a very nice page on experimental tests of Bell's inequality and related issues.

$$\text{http://en.wikipedia.org/wiki/Bell\_test\_experiments}$$

## Problems

1. *The spin observable for an arbitrary direction.*

   The operator $J(\theta, \phi)$ corresponding to spin in the spatial direction $(\theta, \phi)$ in spherical coordinates on the unit sphere in three dimensions is

   $$J(\theta, \phi) = \sin\theta \cos\phi J_x + \sin\theta \sin\phi J_y + \cos\theta J_z.$$

   Consider the spin-1/2 representation.

   (a) Show that the eigenvalues of $J(\theta, \phi)$ are $\pm \hbar/2$ and that the corresponding normalised eigenvectors may be taken to be

   $$|\theta, \phi\rangle = \cos\left(\frac{\theta}{2}\right)\left|\frac{1}{2}\ \frac{1}{2}\right\rangle + e^{i\phi}\sin\left(\frac{\theta}{2}\right)\left|\frac{1}{2}\ -\frac{1}{2}\right\rangle,$$

   and

   $$|\pi - \theta, \phi + \pi\rangle = \sin\left(\frac{\theta}{2}\right)\left|\frac{1}{2}\ \frac{1}{2}\right\rangle - e^{i\phi}\cos\left(\frac{\theta}{2}\right)\left|\frac{1}{2}\ -\frac{1}{2}\right\rangle,$$

   where $\left|\frac{1}{2}\ \pm\frac{1}{2}\right\rangle$ satisfy $J_z \left|\frac{1}{2}\ \pm\frac{1}{2}\right\rangle = \pm\frac{\hbar}{2}\left|\frac{1}{2}\ \pm\frac{1}{2}\right\rangle$.

   (b) Show that the pair of vectors $|\theta, \phi\rangle$ and $|\pi - \theta, \phi + \pi\rangle$ form an orthonormal basis for $\mathbb{C}^2$ (for fixed values of $(\theta, \phi)$).

   (c) Show also that the operator $J(\theta, \phi)$ may be written in terms of the projectors onto $|\theta, \phi\rangle$ and $|\pi - \theta, \phi + \pi\rangle$ as

   $$J(\theta, \phi) = \frac{\hbar}{2}|\theta, \phi\rangle\langle\theta, \phi| - \frac{\hbar}{2}|\pi - \theta, \phi + \pi\rangle\langle\pi - \theta, \phi + \pi|.$$

---

[12] Daniel Garisto. The universe is not locally real, and the physics Nobel Prize winners proved it. *Scientific American*, 6, 2022.

2. *Measurement of spin in arbitrary directions.*

   Consider a spin-1/2 particle in the state $\left|\frac{1}{2}\ \frac{1}{2}\right\rangle$. By writing the state in terms of the eigenstates of $J(\theta,\phi)$ (defined in question 1), or otherwise, calculate the probability that the eigenvalue $\hbar/2$ is found when $J(\theta,\phi)$ is measured.

3. *The uncertainty relations for spin.*

   (a) Derive the following uncertainty relation for spin from the commutation relations:

   $$\Delta_{|\psi\rangle}(J_x)\ \Delta_{|\psi\rangle}(J_y) \geq \frac{\hbar}{2}\left|E_{|\psi\rangle}(J_z)\right|. \tag{5.80}$$

   (note that the right hand side of this equation depends on the state $|\psi\rangle$ unlike the case for the canonical commutation relations). Under what circumstances is there equality in (5.80)?

   (b) Calculate the terms in the uncertainty relation above for the state $|\psi\rangle = \left|\frac{1}{2}\ \frac{1}{2}\right\rangle$, and confirm that the uncertainty relation is satisfied. Comment on your answer in the light of your answer to the last part of (a).

   (c) Calculate $\Delta_{|\psi\rangle}(J_x)$ for the state

   $$|\psi\rangle = \frac{1}{\sqrt{2}}\left(\left|\frac{1}{2}\ \frac{1}{2}\right\rangle + \left|\frac{1}{2}\ -\frac{1}{2}\right\rangle\right).$$

   Is this value consistent with the uncertainty relation?

4. *Tensor product operations.*

   Let us define the operators $X, Y$ and $Z$ on $\mathbb{C}^2$ by

   $$\begin{aligned}
   X\,|1\rangle &= |2\rangle; & X\,|2\rangle &= |1\rangle \\
   Y\,|1\rangle &= i\,|2\rangle; & Y\,|2\rangle &= -i\,|1\rangle \\
   Z\,|1\rangle &= |1\rangle; & Z\,|2\rangle &= -\,|2\rangle.
   \end{aligned}$$

   and let $\mathbf{I}$ denote the identity operator on $\mathbb{C}^2$.

   (a) Show that the operator $X \otimes \mathbf{I}$ on $\mathbb{C}^2 \otimes \mathbb{C}^2$ is unitary. You may use, without proof, the facts that $(A \otimes B)^\dagger = A^\dagger \otimes B^\dagger$, and $(A \otimes B)(C \otimes D) = AC \otimes BD$, for operators $A, B, C, D$.

   (b) Let $|\Psi\rangle$ be the state

   $$|\Psi\rangle = \frac{1}{\sqrt{2}}\left(|1\rangle\,|2\rangle - |2\rangle\,|1\rangle\right)$$

   on $\mathbb{C}^2 \otimes \mathbb{C}^2$. Calculate $|\Psi_X\rangle = X \otimes \mathbf{I}\,|\Psi\rangle$.

(c) Calculate also $|\Psi_Y\rangle = Y \otimes \mathbf{I}\,|\Psi\rangle$, and $|\Psi_Z\rangle = Z \otimes \mathbf{I}\,|\Psi\rangle$. Show that the four states $|\Psi\rangle$, $|\Psi_X\rangle$, $|\Psi_Y\rangle$, $|\Psi_Z\rangle$ form an orthonormal basis for $\mathbb{C}^2 \otimes \mathbb{C}^2$.

5. Suppose we have a source that creates two quantum particles of spin $\frac{1}{2}$. One is sent to Alice and one is sent to Bob ("A" and "B"). The two particles that are sent to Alice and Bob are created by the source through some type of physical process and they form a composite system. Figure 5.6 shows a schematic diagram of the two particles created by the source, with one sent to Alice and one sent to Bob. Alice and Bob can each independently perform measurements on their particles. We assume that Alice and Bob are sufficiently spatially separated that no communication between them is possible in the time that it takes them to perform their measurements (even at the speed of light). Let $\{|0\rangle, |1\rangle\}$ denote an orthonormal basis of $\mathbb{C}^2$ and let $Z$ be the operator on $\mathbb{C}^2$ defined by:

$$Z|0\rangle = |0\rangle, \qquad Z|1\rangle = -1.$$

The Hilbert space for the composite system of two spin $\frac{1}{2}$ particles is $\mathbb{C}^2 \otimes \mathbb{C}^2$, and we assume that the two particles are in the following state:

$$|\psi\rangle = \frac{1}{\sqrt{2}} \left( |0\rangle|0\rangle + |1\rangle|1\rangle \right). \tag{5.81}$$

Let us suppose that Alice measures $Z$, and Bob does not do anything. Then the measurement operator on $\mathbb{C}^2 \otimes \mathbb{C}^2$ is $Z \otimes \mathbb{I}$.

(a) Show that $Z \otimes \mathbb{I}$ is self-adjoint.
(b) Compute the eigenvalues and eigenvectors of $Z \otimes \mathbb{I}$.
(c) Express $Z \otimes \mathbb{I}$ in spectral form in Dirac notation.
(d) Show that $|\psi\rangle$ is entangled.
(e) Suppose the outcome of Alice's measurement is $+1$. Determine the state of the system after this measurement.
(f) Suppose Alice measures $+1$. What is the probability that Bob measures $+1$ after Alice's measurement? What is the probability that Bob measures $-1$ after Alice's measurement?

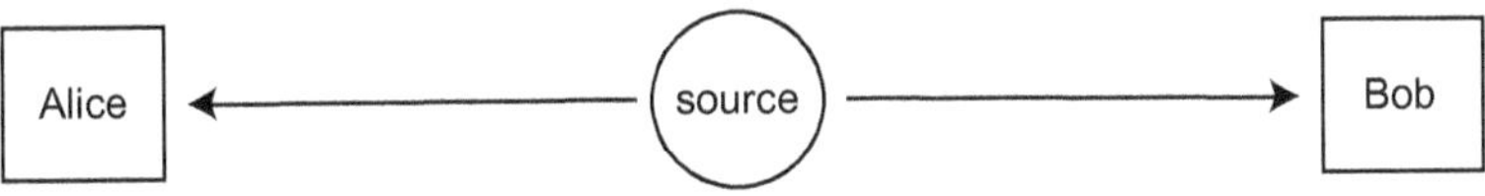


Figure 5.6: Schematic of the source creating two particles, one is sent to Alice and one is sent to Bob.

# Solutions to Problems

## Chapter 1

1. (a)

$$\|\psi + \phi\|^2 = |(\psi + \phi, \psi + \phi)|$$
$$= |(\psi, \psi) + (\psi, \phi) + (\phi, \psi) + (\phi, \phi)|$$
$$\leq |(\psi, \psi)| + |(\psi, \phi)| + |(\phi, \psi)| + |(\phi, \phi)|,$$
$$= \|\psi\|^2 + 2|(\psi, \phi)| + \|\phi\|^2.$$

Applying the Schwarz inequality to the second term:

$$\|\psi + \phi\|^2 \leq \|\psi\|^2 + 2\|\phi\|\|\psi\| + \|\phi\|^2,$$
$$= (\|\psi\| + \|\phi\|)^2.$$

So, $\|\psi + \phi\| \leq \|\psi\| + \|\phi\|$.

(b) $A + B$ is bounded because:

$$\|(A + B)\psi\| = \|A\psi + B\psi\| \leq \|A\psi\| + \|B\psi\|$$
$$\leq a\|\psi\| + b\|\psi\| = (a + b)\|\psi\|.$$

$AB$ is bounded because

$$\|AB\psi\| = \|A(B\psi)\| \leq a\|B\psi\| \leq ab\|\psi\|.$$

2. (a) The matrix representation of the operator $A$ is

$$\begin{pmatrix} 0 & 2 \\ 1 & 1 \end{pmatrix}.$$

Note that since the basis is not orthonormal, the matrix elements are not given by eg. $A_{11} = (e_1, Ae_1)$. Just read them off from $Ae_i = \sum_{j=1,2} A_{ji}e_j$.

The matrix representation of the operator $B$ is

$$\begin{pmatrix} 0 & -i \\ i & 0 \end{pmatrix}.$$

(b)

$$A(Be_1) = A(ie_2) = iAe_2 = 2ie_1 + ie_2,$$
$$A(Be_2) = A(-ie_1) = -iAe_1 = -ie_2.$$

The matrix of representation $AB$ is $\begin{pmatrix} 2i & 0 \\ i & -i \end{pmatrix}.$

(c) Multiplying the matrix representation for $A$ and the matrix representation for $B$:

$$\begin{pmatrix} 0 & 2 \\ 1 & 1 \end{pmatrix} \begin{pmatrix} 0 & -i \\ i & 0 \end{pmatrix} = \begin{pmatrix} 2i & 0 \\ i & -i \end{pmatrix}.$$

(d) (i)

$$A\phi = A(2e_1 + 3ie_2),$$
$$= 2Ae_1 + 3iAe_2,$$
$$= 2e_2 + 3i(2e_1 + e_2),$$
$$= 6ie_1 + (2 + 3i)e_2.$$

Thus

$$(\phi, A\phi) = (2e_1 + 3ie_2, 6ie_1 + (2 + 3i)e_2),$$
$$= (2e_1, 6ie_1) + (2e_1, (2 + 3i)e_2) + (3ie_2, 6ie_1)$$
$$+ (3ie_2, (2 + 3i)e_2),$$
$$= 12i(e_1, e_1) + (4 + 6i)(e_1, e_2)$$
$$+ (-3i)(6i)(e_2, e_1) + (-3i)(2 + 3i)(e_2, e_2),$$
$$= 12i - 6i + 9 = 9 + 6i.$$

(ii) Now using matrix notation:

$$\begin{pmatrix} 2 & -3i \end{pmatrix} \begin{pmatrix} 0 & 2 \\ 1 & 1 \end{pmatrix} \begin{pmatrix} 2 \\ 3i \end{pmatrix} = \begin{pmatrix} 2 & -3i \end{pmatrix} \begin{pmatrix} 6i \\ 2 + 3i \end{pmatrix},$$
$$= 12i - 6i + 9 = 9 + 6i.$$

3. Recall that the adjoint $A^\dagger$ of $A$ is defined by

$$(\phi, A^\dagger \psi) = (A\phi, \psi),$$

for all vectors $\phi$, $\psi$ in the given Hilbert space.

(a)

$$
\begin{aligned}
(\phi, (A^\dagger)^\dagger \psi) &= (A^\dagger \phi, \psi) \\
&= \overline{(\psi, A^\dagger \phi)} \\
&= \overline{(A\psi, \phi)} \\
&= (\phi, A\psi) \\
\Rightarrow \quad A &= (A^\dagger)^\dagger
\end{aligned}
$$

(b)

$$
\begin{aligned}
(\phi, (A+B)^\dagger \psi) &= ((A+B)\phi, \psi) \\
&= (A\phi, \psi) + (B\phi, \psi) \\
&= (\phi, A^\dagger \psi) + (\phi, B^\dagger \psi) \\
&= (\phi, (A^\dagger + B^\dagger)\psi) \\
\Rightarrow \quad (A+B)^\dagger &= (A^\dagger + B^\dagger)
\end{aligned}
$$

(c)

$$
\begin{aligned}
(\phi, (\alpha A)^\dagger \psi) &= (\alpha A\phi, \psi) \\
&= \bar\alpha (A\phi, \psi) \\
&= \bar\alpha (\phi, A^\dagger \psi) \\
&= (\phi, \bar\alpha A^\dagger \psi) \\
\Rightarrow \quad (\alpha A)^\dagger &= \bar\alpha A^\dagger
\end{aligned}
$$

(d)

$$
\begin{aligned}
(\phi, (AB)^\dagger \psi) &= (AB\phi, \psi) \\
&= (A(B\phi), \psi) \\
&= (B\phi, A^\dagger \psi) \\
&= (\phi, B^\dagger A^\dagger \psi) \\
\Rightarrow \quad (AB)^\dagger &= B^\dagger A^\dagger
\end{aligned}
$$

(e) For any vectors $\phi$ and $\psi$

$$(\phi, (A^{-1})^\dagger A^\dagger \psi) = (A^{-1}\phi, A^\dagger \psi)$$
$$= (AA^{-1}\phi, \psi) = (\phi, \psi)$$

and

$$(\phi, A^\dagger (A^{-1})^\dagger \psi) = (A\phi, (A^{-1})^\dagger \psi) = (A^{-1}A\phi, \psi) = (\phi, \psi)$$

so

$$(A^{-1})^\dagger A^\dagger = \mathbb{1} = A^\dagger (A^{-1})^\dagger,$$

where $\mathbb{1}$ denotes the identity operator.

(Remember that $A$ and $B$ are inverses of each other if $AB = \mathbb{1} = BA$.)

4. (a) Let $|\eta\rangle$ and $|\chi\rangle$ be vectors in the vector space acted on by the operator $|\psi\rangle\langle\phi|$. If $A = |\psi\rangle\langle\phi|$, then let $A^\dagger$ be the adjoint. Since $\langle\eta|A$ is the bra-vector corresponding to $A^\dagger|\eta\rangle$, we have

$$\langle\chi|(A^\dagger|\eta\rangle) = \overline{(\langle\eta|A)|\chi\rangle}$$
$$= \overline{\langle\eta|\psi\rangle\langle\phi|\chi\rangle}$$
$$= \langle\chi|\phi\rangle\langle\psi|\eta\rangle.$$

so $A^\dagger = |\phi\rangle\langle\psi|$.

(b) $P_k = \sum_{i=1}^{k} |e_i\rangle\langle e_i|$ so

$$P_k^\dagger = \sum_{i=1}^{k} (|e_i\rangle\langle e_i|)^\dagger = \sum_{i=1}^{k} |e_i\rangle\langle e_i|,$$

using part (a).

Also, since the $|e_i\rangle$ are orthonormal,

$$P_k^2 = \sum_{i=1}^{k}\sum_{j=1}^{k} |e_i\rangle\langle e_i|e_j\rangle\langle e_j|$$
$$= \sum_{i=1}^{k}\sum_{j=1}^{k} |e_i\rangle\delta_{ij}\langle e_j| \quad \text{so } j = i$$
$$= \sum_{i=1}^{k} |e_i\rangle\langle e_i| = P_k.$$

(c) Let $|\phi\rangle$ be an eigenstate of $P_k$, so $P_k|\phi\rangle = \rho|\phi\rangle$. Acting on both sides with $P_k$,

$$P_k^2|\phi\rangle = \rho P_k|\phi\rangle = \rho^2|\phi\rangle$$

but $P_k^2 = P_k$ so

$$P_k|\phi\rangle = \rho^2|\phi\rangle \Rightarrow \rho|\phi\rangle = \rho^2|\phi\rangle \Rightarrow \rho^2 = \rho.$$

So $\rho = 0$ or $1$.

5. (a)

$$H = \begin{pmatrix} -1/2 & \sqrt{3}/2 \\ \sqrt{3}/2 & 1/2 \end{pmatrix}$$

$H^\dagger$ is the conjugate transpose but since the off-diagonal entries are the same and real, $H = H^\dagger$.

(b) We can write $H$ as

$$H = -\frac{1}{2}|1\rangle\langle 1| + \frac{\sqrt{3}}{2}|1\rangle\langle 2| + \frac{\sqrt{3}}{2}|2\rangle\langle 1| + \frac{1}{2}|2\rangle\langle 2|,$$

so

$$H|1\rangle = -\frac{1}{2}|1\rangle\langle 1|1\rangle + \frac{\sqrt{3}}{2}|1\rangle\langle 2|1\rangle$$

$$+ \frac{\sqrt{3}}{2}|2\rangle\langle 1|1\rangle + \frac{1}{2}|2\rangle\langle 2|1\rangle$$

$$= -\frac{1}{2}|1\rangle + \frac{\sqrt{3}}{2}|2\rangle$$

and

$$H|2\rangle = -\frac{1}{2}|1\rangle\langle 1|2\rangle + \frac{\sqrt{3}}{2}|1\rangle\langle 2|2\rangle$$

$$+ \frac{\sqrt{3}}{2}|2\rangle\langle 1|2\rangle + \frac{1}{2}|2\rangle\langle 2|2\rangle$$

$$= \frac{\sqrt{3}}{2}|1\rangle + \frac{1}{2}|2\rangle.$$

(c) We check that $H^2 = I$:

$$H^2 = \begin{pmatrix} -1/2 & \sqrt{3}/2 \\ \sqrt{3}/2 & 1/2 \end{pmatrix} \begin{pmatrix} -1/2 & \sqrt{3}/2 \\ \sqrt{3}/2 & 1/2 \end{pmatrix}$$

$$= \begin{pmatrix} 1 & 0 \\ 0 & 1 \end{pmatrix} = I.$$

(d) To calculate the eigenvalues of $H$, we let

$$\det \begin{pmatrix} -1/2 - \lambda & \sqrt{3}/2 \\ \sqrt{3}/2 & 1/2 - \lambda \end{pmatrix} = 0$$

$$(-1/2 - \lambda)(1/2 - \lambda) - 3/4 = 0$$

$$\lambda^2 = 1$$

$$\lambda = \pm 1$$

If $\lambda = 1$ then we calculate the corresponding eigenvector:

$$\begin{pmatrix} -3/2 & \sqrt{3}/2 \\ \sqrt{3}/2 & -1/2 \end{pmatrix} \begin{pmatrix} a \\ b \end{pmatrix} = \begin{pmatrix} 0 \\ 0 \end{pmatrix}.$$

This gives us that $a = b/\sqrt{3}$. The eigenvector is therefore some multiple of $(\frac{1}{\sqrt{3}}, 1)$. We normalise it to

$$|e_1\rangle = \tfrac{1}{2}|1\rangle + \tfrac{\sqrt{3}}{2}|2\rangle.$$

If $\lambda = -1$, then calculating the corresponding eigenvector:

$$\begin{pmatrix} 1/2 & \sqrt{3}/2 \\ \sqrt{3}/2 & 3/2 \end{pmatrix} \begin{pmatrix} a \\ b \end{pmatrix} = \begin{pmatrix} 0 \\ 0 \end{pmatrix}.$$

So $a = -\sqrt{3}b$ and the eigenvector is some multiple of $(-\sqrt{3}, 1)$. We normalise it to

$$|e_2\rangle = \tfrac{-\sqrt{3}}{2}|1\rangle + \tfrac{1}{2}|2\rangle.$$

(e) We can now write $H$ in Dirac notation in terms of the basis of orthonormal eigenvectors. By plugging in the expressions for each of the eigenvectors we can retrieve the result of question 1:

$$\begin{aligned}
H &= |e_1\rangle\langle e_1| - |e_2\rangle\langle e_2| \\
&= (\tfrac{1}{2}|1\rangle + \tfrac{\sqrt{3}}{2}|2\rangle)(\tfrac{1}{2}\langle 1| + \tfrac{\sqrt{3}}{2}\langle 2|) \\
&\quad - (-\tfrac{\sqrt{3}}{2}|1\rangle + \tfrac{1}{2}|2\rangle)(-\tfrac{\sqrt{3}}{2}\langle 1| + \tfrac{1}{2}\langle 2|) \\
&= \tfrac{1}{4}|1\rangle\langle 1| + \tfrac{\sqrt{3}}{4}|1\rangle\langle 2| + \tfrac{\sqrt{3}}{4}|2\rangle\langle 1| + \tfrac{3}{4}|2\rangle\langle 2| - \tfrac{3}{4}|1\rangle\langle 1| \\
&\quad + \tfrac{\sqrt{3}}{4}|1\rangle\langle 2| + \tfrac{\sqrt{3}}{4}|2\rangle\langle 1| - \tfrac{1}{4}|2\rangle\langle 2| \\
&= -\tfrac{1}{2}|1\rangle\langle 1| + \tfrac{\sqrt{3}}{2}|1\rangle\langle 2| + \tfrac{\sqrt{3}}{2}|2\rangle\langle 1| + \tfrac{1}{2}|2\rangle\langle 2|.
\end{aligned}$$

6. (a) Since $H^2 = I$, we have $H^{2k} = I$ and $H^{2k+1} = H$ (for integer $k$). Therefore

$$
\begin{aligned}
e^{-itH} &= \sum_{n=0}^{\infty} \frac{(-it)^n H^n}{n!} \\
&= \sum_{k=0}^{\infty} \frac{(-it)^{2k} H^{2k}}{(2k)!} + \sum_{k=0}^{\infty} \frac{(-it)^{2k+1} H^{2k+1}}{(2k+1)!} \\
&= \sum_{k=0}^{\infty} \frac{(-1)^k t^{2k}}{(2k)!} I + \sum_{k=0}^{\infty} \frac{(-1)^k (-i) t^{2k+1}}{(2k+1)!} H \\
&= \cos t\, I - i \sin t\, H.
\end{aligned}
$$

(b) In general, we can show that $e^{-itH}$ is unitary. Let $U = e^{-itH}$. Then

$$
\begin{aligned}
UU^\dagger &= (e^{-itH})(e^{-itH})^\dagger \\
&= (e^{-itH})(e^{itH^\dagger}), \quad \text{but } H = H^\dagger \\
&= (e^{-itH})(e^{itH}).
\end{aligned}
$$

But $e^A e^{-A} = I$ for operator $A$, just as $e^\theta e^{-\theta} = 1$ for $\theta \in \mathbb{C}$. We can see this because

$$
\begin{aligned}
1 &= e^\theta e^{-\theta} \\
&= \sum_{n=0}^{\infty} \frac{\theta^n}{n!} \sum_{m=0}^{\infty} \frac{(-\theta)^m}{m!} \\
&= \sum_{n,m=0}^{\infty} \frac{\theta^n (-\theta)^m}{n!\,m!};
\end{aligned}
$$

that is, the coefficient of the $\theta^r$ term is zero for $r > 0$.
Since

$$
e^A e^{-A} = \sum_{n=0}^{\infty} \frac{A^n}{n!} \sum_{m=0}^{\infty} \frac{(-A)^m}{m!},
$$

the coefficient of $A^r$, $r > 0$, is again zero since it makes no difference here that $A$ is an operator.
So, $UU^\dagger = (e^{-itH})(e^{itH}) = I$.

For the particular Hamiltonian in this question that satisfies $H^2 = I$, we have:

$$(e^{-itH})(e^{-itH})^\dagger = (\cos t\, I - i\sin t\, H)(\cos t\, I + i\sin t\, H)$$

$$= \cos^2 t\, I + i\sin t\,\cos t\, H$$

$$- i\sin t\,\cos t\, H + \sin^2 t\, H^2,$$

but $H^2 = I$, so

$$(e^{-itH})(e^{-itH})^\dagger = I(\sin^2 t + \cos^2 t) = I.$$

(c) We have $|\psi(t)\rangle = e^{-itH}|1\rangle$.

Remember that

$$|e_1\rangle = \tfrac{1}{2}|1\rangle + \tfrac{\sqrt{3}}{2}|2\rangle$$

and

$$|e_2\rangle = -\tfrac{\sqrt{3}}{2}|1\rangle + \tfrac{1}{2}|2\rangle.$$

If we multiply the second equation by $-\sqrt{3}$ and add the two, then we find

$$|1\rangle = \tfrac{1}{2}|e_1\rangle - \tfrac{\sqrt{3}}{2}|e_2\rangle.$$

So,

$$|\psi(t)\rangle = e^{-itH}\left(\tfrac{1}{2}|e_1\rangle - \tfrac{\sqrt{3}}{2}|e_2\rangle\right)$$

$$= \tfrac{1}{2}e^{-it}|e_1\rangle - \tfrac{\sqrt{3}}{2}e^{it}|e_2\rangle.$$

Now

$$H|\psi(t)\rangle = \tfrac{1}{2}e^{-it}|e_1\rangle + \tfrac{\sqrt{3}}{2}e^{it}|e_2\rangle,$$

and

$$i\frac{\partial}{\partial t}|\psi(t)\rangle = \frac{i}{2}(-i)e^{-it}|e_1\rangle - \frac{\sqrt{3}i}{2}(i)e^{it}|e_2\rangle$$

$$= \frac{1}{2}e^{-it}|e_1\rangle + \frac{\sqrt{3}}{2}e^{it}|e_2\rangle,$$

so

$$i\frac{\partial}{\partial t}|\psi(t)\rangle = H|\psi(t)\rangle$$

and Schrödinger's equation is satisfied

7. (a) With

$$H = \begin{pmatrix} 3 & 1 \\ 1 & 3 \end{pmatrix},$$

it is clear that $H = H^\dagger$.

(b)

$$H = 3 \mid v_1\rangle\langle v_1 \mid +1 \mid v_1\rangle\langle v_2 \mid +1 \mid v_2\rangle\langle v_1 \mid +3 \mid v_2\rangle\langle v_2 \mid$$

(c) The expectation value of $H$ in the state $\mid v_1\rangle$ is given by:

$$\langle v_1 \mid H \mid v_1 \rangle = 3.$$

(d) Recall

$$H = \begin{pmatrix} 3 & 1 \\ 1 & 3 \end{pmatrix}.$$

We compute the eigenvalues of $H$ by solving for the roots of the characteristic equation:

$$\det(H - E\,\mathbb{I}) = \det \begin{pmatrix} 3 - E & 1 \\ 1 & 3 - E \end{pmatrix} = E^2 - 6E + 8 = 0.$$

We denote the two roots (eigenvalues) by:

$$E_1 = 2, \quad E_2 = 4.$$

Next we compute the eigenvectors corresponding to each eigenvalue.

The eigenvector corresponding to $E_1 = 2$ obtained as follows:

$$\begin{pmatrix} 3 & 1 \\ 1 & 3 \end{pmatrix} \begin{pmatrix} x \\ y \end{pmatrix} = 2 \begin{pmatrix} x \\ y \end{pmatrix}.$$

This gives the condition:

$$x = -y,$$

from which we obtain:

$$\begin{pmatrix} 1 \\ -1 \end{pmatrix}$$

Normalizing this eigenvector gives:

$$\frac{1}{\sqrt{2}} \begin{pmatrix} 1 \\ -1 \end{pmatrix} \equiv \mid E_1\rangle.$$

Similarly, the eigenvector corresponding to $E_2 = 4$ is obtained from the following calculation:

$$\begin{pmatrix} 3 & 1 \\ 1 & 3 \end{pmatrix} \begin{pmatrix} x \\ y \end{pmatrix} = 4 \begin{pmatrix} x \\ y \end{pmatrix},$$

which gives the condition:

$$x = y,$$

from which we obtain the eigenvector:

$$\begin{pmatrix} 1 \\ 1 \end{pmatrix}.$$

normalizing this eigenvector gives:

$$\frac{1}{\sqrt{2}} \begin{pmatrix} 1 \\ 1 \end{pmatrix} \equiv |\, E_2 \rangle.$$

(e)

$$H = 2 \,|\, E_1 \rangle\langle E_1 \,| + 4 \,|\, E_2 \rangle\langle E_2 \,|.$$

(f)

$$e^{-\frac{iHt}{\hbar}} = e^{-\frac{iE_1 t}{\hbar}} \,|\, E_1 \rangle\langle E_1 \,| + e^{-\frac{iE_2 t}{\hbar}} \,|\, E_2 \rangle\langle E_2 \,|.$$

8. (a) We want to determine the bra associated with the ket $A \,|\, \psi \rangle$. Towards this end, we let $|\, u \rangle = A \,|\, \psi \rangle$, and then we perform the following calculation:

$$\langle u \,|\, v \rangle = \overline{\langle v \,|\, u \rangle},$$
$$= \overline{\langle v \,|\, A \,|\, \psi \rangle},$$
$$= \langle \psi \,|\, A^\dagger \,|\, v \rangle, \tag{1.1}$$

which is true for any $|\, v \rangle$. Hence, we have:

$$\langle u \,| = (A \,|\, \psi \rangle)^\dagger = \langle \psi \,|\, A^\dagger. \tag{1.2}$$

(b)   (i) scalar,    $\overline{\langle \psi | A | \phi \rangle}\, \overline{\langle \psi | \phi \rangle}$

     (ii) bra,    $\overline{\langle \psi | \phi \rangle}\, A^\dagger | \psi \rangle$

     (iii) operator,    $\overline{\langle \psi | \phi \rangle}\, | \psi \rangle \langle \phi | A^\dagger$

     (iv) ket,    $\overline{\langle \phi | A | \psi \rangle}\, \langle \psi | A^\dagger$

(c) Let $|\,\mathbf{e}\,\rangle$ be a normalised eigenvector of $A$ with eigenvalues $\lambda$, i.e.,

$$A\,|\,\mathbf{e}\,\rangle = \lambda\,|\,\mathbf{e}\,\rangle, \quad \langle\,\mathbf{e}\,|\,\mathbf{e}\,\rangle = 1.$$

Then we have:

$$\begin{aligned}
\lambda = \langle\,\mathbf{e}\,|\,A\,|\,\mathbf{e}\,\rangle &= \overline{\langle\,\mathbf{e}\,|\,A^\dagger\,|\,\mathbf{e}\,\rangle}, \\
&= \overline{\langle\,\mathbf{e}\,|\,A\,|\,\mathbf{e}\,\rangle}, \quad \text{since} \quad A = A^\dagger, \\
&= \bar{\lambda}\overline{\langle\,\mathbf{e}\,|\,\mathbf{e}\,\rangle}, \\
&= \bar{\lambda}, \hspace{4cm} (1.3)
\end{aligned}$$

from which it follows that $\lambda$ is real.

(d) We consider eigenvectors $|\,\mathbf{e}_1\,\rangle$, $|\,\mathbf{e}_2\,\rangle$ such that:

$$\begin{aligned}
A\,|\,\mathbf{e}_1\,\rangle &= \lambda_1\,|\,\mathbf{e}_1\,\rangle, \\
A\,|\,\mathbf{e}_2\,\rangle &= \lambda_2\,|\,\mathbf{e}_2\,\rangle.
\end{aligned} \hspace{2cm} (1.4)$$

Then we have:

$$\begin{aligned}
\lambda_1\langle\,\mathbf{e}_2\,|\,\mathbf{e}_1\,\rangle &= \langle\,\mathbf{e}_2\,|\,A\,|\,\mathbf{e}_1\,\rangle \\
&= \overline{\langle\,\mathbf{e}_1\,|\,A^\dagger\,|\,\mathbf{e}_2\,\rangle}, \\
&= \overline{\langle\,\mathbf{e}_1\,|\,A\,|\,\mathbf{e}_2\,\rangle}, \quad \text{since} \quad A = A^\dagger, \\
&= \bar{\lambda}_2\overline{\langle\,\mathbf{e}_1\,|\,\mathbf{e}_2\,\rangle}, \\
&= \lambda_2\langle\,\mathbf{e}_2\,|\,\mathbf{e}_1\,\rangle. \hspace{2cm} (1.5)
\end{aligned}$$

Hence, it follows that:

$$(\lambda_1 - \lambda_2)\langle\,\mathbf{e}_2\,|\,\mathbf{e}_1\,\rangle = 0 \quad \Rightarrow \quad \langle\,\mathbf{e}_2\,|\,\mathbf{e}_1\,\rangle = 0.$$

9. (a)

$$A = \begin{pmatrix} 2 & i \\ -i & 2 \end{pmatrix}, \quad B = \begin{pmatrix} 1 & i \\ -i & 1 \end{pmatrix}$$

(b) It is trivial to verify that $A = A^*$ and $B = B^*$.

(c) We have

$$\begin{aligned}
AB - BA &= \begin{pmatrix} 2 & i \\ -i & 2 \end{pmatrix}\begin{pmatrix} 1 & i \\ -i & 1 \end{pmatrix} - \begin{pmatrix} 1 & i \\ -i & 1 \end{pmatrix}\begin{pmatrix} 2 & i \\ -i & 2 \end{pmatrix} \\
&= \begin{pmatrix} 0 & 0 \\ 0 & 0 \end{pmatrix}.
\end{aligned}$$

(d)

$$\det\left(A - \lambda\mathbb{I}\right) = (2 - \lambda)^2 - 1$$

$$= \lambda^2 - 4\lambda + 3 = 0, \Rightarrow \lambda = 1, 3.$$

For $\lambda = 1$ the corresponding normalized eigenvector is computed as follows:

$$\begin{pmatrix} 2 & i \\ -i & 2 \end{pmatrix} \begin{pmatrix} a \\ b \end{pmatrix} = \begin{pmatrix} a \\ b \end{pmatrix}$$

which implies:

$$a + ib = 0$$

giving rise to the eigenvector:

$$\begin{pmatrix} a \\ b \end{pmatrix} = \begin{pmatrix} 1 \\ i \end{pmatrix},$$

which after normalization gives:

$$|e_1\rangle = \frac{1}{\sqrt{2}} \begin{pmatrix} 1 \\ i \end{pmatrix}.$$

For $\lambda = 3$ the corresponding normalized eigenvector is computed as follows:

$$\begin{pmatrix} 2 & i \\ -i & 2 \end{pmatrix} \begin{pmatrix} a \\ b \end{pmatrix} = 3 \begin{pmatrix} a \\ b \end{pmatrix}$$

which implies:

$$-a + ib = 0,$$

giving rise to the eigenvector:

$$\begin{pmatrix} a \\ b \end{pmatrix} = \begin{pmatrix} -1 \\ i \end{pmatrix},$$

which after normalization gives:

$$|e_2\rangle = \frac{1}{\sqrt{2}} \begin{pmatrix} -1 \\ i \end{pmatrix}.$$

(e) It is straightforward to verify:

$$|1\rangle = \frac{1}{\sqrt{2}}|e_1\rangle - \frac{1}{\sqrt{2}}|e_2\rangle,$$

$$|2\rangle = \frac{1}{i\sqrt{2}}|e_1\rangle + \frac{1}{i\sqrt{2}}|e_2\rangle.$$

(f) The eigenvalues of $B$ are given by:
$$\det\left(B - \lambda\mathbb{I}\right) = (1-\lambda)^2 - 1 = \lambda^2 - 2\lambda = 0, \Rightarrow \lambda = 0,\ 2.$$

It is easy to verify that the normalized eigenvector corresponding to the eigenvalue 2 is $|e_2\rangle$. Therefore an orthonormal eigenvector corresponding to the eigenvalue 0 is $|e_1\rangle$.

(g)
$$B = 2|e_2\rangle\langle e_2|.$$

(h)
$$\langle e_2|B|e_2\rangle = 2\langle e_2|e_2\rangle\langle e_2|e_2\rangle = 2.$$

10. (a)
$$\begin{aligned}
\mathcal{O}(\alpha\psi + \beta\phi) &= x^2(\alpha\psi + \beta\phi), \\
&= \alpha x^2\psi + \beta x^2\phi, \\
&= \alpha\mathcal{O}\psi + \beta\mathcal{O}\phi, \quad \forall\alpha,\ \beta \in \mathbb{C}.
\end{aligned}$$

(b)
$$\begin{aligned}
\mathcal{N}(\alpha\psi + \beta\phi) &= \int_0^x (\alpha\psi(x') + \beta\phi(x'))dx', \\
&= \alpha\int_0^x \psi(x')dx' + \beta\int_0^x \phi(x')dx', \\
&= \alpha\mathcal{N}\psi + \beta\mathcal{N}\phi, \quad \forall\alpha,\ \beta \in \mathbb{C}.
\end{aligned}$$

(c)
$$\int \overline{\mathcal{O}\psi(x)}\,\phi(x)dx = \int \overline{\psi(x)}\mathcal{O}\phi(x)dx.$$

$$\begin{aligned}
\int \overline{\mathcal{O}\psi(x)}\,\phi(x)dx &= \int x^2\overline{\psi(x)}\phi(x)dx, \\
&= \int \overline{\psi(x)}x^2\phi(x), \\
&= \int \overline{\psi(x)}\mathcal{O}\phi(x)dx
\end{aligned}$$

(d)
$$\mathcal{O}\mathcal{N}\psi(x) = x^2\int_0^x \psi(x')dx'$$

$$\mathcal{N}\mathcal{O}\psi(x) = \int_0^x x'^2\psi(x')dx'$$

11. (a) We use the spectral representation of self-adjoint operators on complex inner product spaces in Dirac notation:

$$A = 7 \mid e_1 \rangle \langle e_1 \mid + 3 \mid e_2 \rangle \langle e_2 \mid .$$

(b)

$$A = \frac{7}{2}\begin{pmatrix} 1 \\ -1 \end{pmatrix}(1 \ -1) + \frac{3}{2}\begin{pmatrix} 1 \\ 1 \end{pmatrix}(1 \ 1),$$

$$= \frac{7}{2}\begin{pmatrix} 1 & -1 \\ -1 & 1 \end{pmatrix} + \frac{3}{2}\begin{pmatrix} 1 & 1 \\ 1 & 1 \end{pmatrix},$$

$$= \begin{pmatrix} 5 & -2 \\ -2 & 5 \end{pmatrix}.$$

(c) This is impossible since self-adjoint operators have real eigenvalues.

## Chapter 2

1. Using elementary trigonometric identities, the initial condition can be written as:

$$\psi(x,0) = \frac{1}{\sqrt{a}}\sin\frac{\pi x}{a} + \frac{1}{\sqrt{a}}\sin\frac{2\pi x}{a}, \tag{2.1}$$

from which it is clear that it is a linear superposition of the first two energy eigenstates for the square well, $\psi_1(x)$ and $\psi_2(x)$:

$$\psi(x,0) = \frac{1}{\sqrt{2}}\psi_1(x) + \frac{1}{\sqrt{2}}\psi_2(x). \tag{2.2}$$

The wavefunction therefore has the form:

$$\psi(x,t) = \frac{1}{\sqrt{2}}\psi_1(x)\exp\left(-\frac{iE_1 t}{\hbar}\right) + \frac{1}{\sqrt{2}}\psi_2(x)\exp\left(-\frac{iE_2 t}{\hbar}\right), \tag{2.3}$$

where, recall, the eigenvalues and eigenstates of the square well are given by:

$$E_n = \frac{n^2\pi^2\hbar^2}{2ma^2}, \qquad \psi_n(x) = \sqrt{\frac{2}{a}}\sin\frac{n\pi x}{a}. \tag{2.4}$$

The probability that at time $t$ the particle lies in the interval $[0, a/2]$ is given by:

$$\int_0^{\frac{a}{2}} |\psi(x,t)|^2 dx = \int_0^{\frac{a}{2}} \left( \frac{1}{2}\psi_1^2(x) + \frac{1}{2}\psi_2^2(x) + \psi_1(x)\psi_2(x) \right.$$

$$\left. \times \cos\left( \frac{E_1 - E_2}{\hbar}t \right) \right) dx,$$

$$= \frac{1}{4} + \frac{1}{4} + \frac{2}{a} \cos\left( \frac{E_1 - E_2}{\hbar}t \right)$$

$$\times \int_0^{\frac{a}{2}} \sin\frac{\pi x}{a} \sin\frac{2\pi x}{a} dx,$$

$$= \frac{1}{2} + \frac{4}{3\pi} \cos\left( \frac{3\pi^2 \hbar t}{2ma^2} \right). \tag{2.5}$$

The probability of finding the particle in the $n^{\text{th}}$ eigenstate is $\frac{1}{2}$ for $n = 1$, 2 and 0 for $n > 2$.

2. We first verify normalization:

$$\int_0^a (\psi(x,0))^2 \, dx = \frac{12}{a^3} \left( \int_0^{\frac{a}{2}} x^2 dx + \int_{\frac{a}{2}}^a (a - x)^2 dx \right) = 1. \tag{2.6}$$

Following the spirit of the previous exercise, we express the initial condition in terms of a linear superposition of the eigenstates of the square well:

$$\psi(x,0) = \sum_{n=1}^{\infty} c_n \psi_n(x). \tag{2.7}$$

The expansion coefficients are given by:

$$c_n = \sqrt{\frac{2}{a}} \int_0^a \psi(x,0) \sin\frac{n\pi x}{a} dx. \tag{2.8}$$

By symmetry $c_n = 0$ if $n$ is even, and

$$c_{2k+1} = 2\sqrt{\frac{2}{a}} \sqrt{\frac{12}{a^3}} \int_0^{\frac{a}{2}} x \sin\frac{(2k+1)\pi x}{a} dx,$$

$$= \frac{4\sqrt{6}}{\pi^2} \frac{(-1)^k}{(2k+1)^2}, \quad k = 0, 1, 2, 3, \ldots \tag{2.9}$$

Therefore the wavefunction has the form:

$$\psi(x,t) = \sum_{k=0}^{\infty} c_{2k+1} \psi_{2k+1}(x) \exp\left(-\frac{E_{2k+1}t}{\hbar}\right). \qquad (2.10)$$

The probability of finding the particle in the interval $[0, \frac{a}{2}] = \frac{1}{2}$. This can be concluded by computing $\int_0^{\frac{a}{2}} |\psi(x,t)|^2 dx$, or by symmetry since the initial condition is symmetric about $\frac{a}{2}$. The probability of finding the particle in the $n^{\text{th}}$ eigenstate is 0 for $n = 2k$ and $|c_{2k+1}|^2$ for $n = 2k + 1$.

3. In terms of the square well $[0, 2a]$ the initial condition has the form:

$$\psi(x,0) = \begin{cases} \sqrt{\frac{2}{a}} \sin \frac{\pi x}{a}, & 0 < x < a, \\ 0, & a \le x < 2a. \end{cases} \qquad (2.11)$$

We expand the initial condition in terms of the energy eigenstates of the square well on the interval $[0, 2a]$, $\phi_n(x) = \frac{1}{\sqrt{a}} \sin \frac{n\pi x}{2a}$ for the interval $[0, 2a]$, with energies $E_n = \frac{n^2 \pi^2 \hbar^2}{8ma^2}$ (replace $a$ by $2a$ in the previously derived formula for the square well on $[0, a]$). The expansion has the form:

$$\psi(x,0) = \sum_{n=1}^{\infty} c_n \phi_n(x), \qquad (2.12)$$

with coefficients given by:

$$c_n = \frac{\sqrt{2}}{a} \int_0^a \sin \frac{\pi x}{a} \sin \frac{n\pi x}{2a} dx. \qquad (2.13)$$

From this expression we obtain $c_2 = \frac{1}{\sqrt{2}}$, $c_{2k} = 0$ for $k > 1$, and

$$c_{2k+1} = \frac{4\sqrt{2}}{\pi}(-1)^{k+1} \frac{1}{(2k-1)(2k+3)}, \quad k = 0, 1, 2, \ldots \qquad (2.14)$$

The probability for the energy to be unchanged is $|c_2|^2 = \frac{1}{2}$. The wavefunction is given by:

$$\psi(x,t) = \sum_{n=1}^{\infty} c_n \phi_n(x) \exp\left(-\frac{i}{\hbar} \frac{n^2 \pi^2 \hbar^2 t}{8ma^2}\right). \qquad (2.15)$$

4. We argue by contradiction. Let $\psi_n$ be an eigenstate with eigenvalue $E_n$ and $\psi_m$ be a *different* eigenstate with eigenvalue $E_m$, and we assume

that $E_n = E_m = E$. In this case we have

$$-\frac{\hbar^2}{2m}\frac{d^2\psi_n}{dx^2} + V(x)\psi_n = E\psi_n,$$

$$-\frac{\hbar^2}{2m}\frac{d^2\psi_m}{dx^2} + V(x)\psi_m = E\psi_m. \tag{2.16}$$

Multiplying the first equation by $\psi_m$, the second by $\psi_n$ and subtracting gives:

$$\psi_m\psi_n'' - \psi_n\psi_m'' = 0, \tag{2.17}$$

which can be rewritten as:

$$\frac{d}{dx}\left(\psi_m\psi_n' - \psi_n\psi_m'\right) = 0, \tag{2.18}$$

from which it follows that:

$$\psi_m\psi_n' - \psi_n\psi_m' = \text{constant}. \tag{2.19}$$

Now since $\psi_n \to 0$, $\psi_m \to 0$ as $|x| \to \infty$ it follows that constant $= 0$, which further implies that:

$$\psi_m\psi_n' - \psi_n\psi_m' = 0, \tag{2.20}$$

or

$$\frac{d}{dx}\log\psi_n = \frac{d}{dx}\log\psi_m. \tag{2.21}$$

Integrating this expression gives:

$$\psi_n = \text{constant }\psi_m. \tag{2.22}$$

This implies that $\psi_n$ and $\psi_m$ are the same state. But this a contradiction that comes from the assumption that $E_n = E_m$.

5. We have

$$\int_{-\infty}^{\infty} \bar{\alpha}\left(-\frac{\hbar^2}{2m}\frac{d^2}{dx^2} + V(x)\right)\beta dx$$

$$= \underbrace{-\frac{\hbar^2}{2m}\int_{-\infty}^{\infty} \bar{\alpha}\frac{d^2\beta}{dx^2}dx}_{I_1} + \underbrace{\int_{-\infty}^{\infty} \bar{\alpha}V(x)\beta dx}_{I_2}, \tag{2.23}$$

and we will compute $I_1$ and $I_2$ separately.

We compute $I_1$ by integrating by parts twice:

$$I_1 = -\frac{\hbar^2}{2m} \left\{ \left[ \frac{d\beta}{dx}\bar{\alpha} \right]_{-\infty}^{\infty} - \int_{-\infty}^{\infty} \frac{d\bar{\alpha}}{dx}\frac{d\beta}{dx} dx \right\},$$

the first term is zero since $\bar{\alpha}(\infty) = \bar{\alpha}(-\infty) = 0$,

$$= \frac{\hbar^2}{2m} \left\{ \left[ \beta\frac{d\bar{\alpha}}{dx} \right]_{-\infty}^{\infty} - \int_{-\infty}^{\infty} \beta\frac{d^2\bar{\alpha}}{dx^2} dx \right\},$$

the first term is zero as above,

$$= -\frac{\hbar^2}{2m} \int_{-\infty}^{\infty} \overline{\beta\frac{d^2\alpha}{dx^2}} dx = -\frac{\hbar^2}{2m} \int_{-\infty}^{\infty} \beta\frac{d^2\bar{\alpha}}{dx^2} dx \tag{2.24}$$

We also have:

$$I_2 = \int_{-\infty}^{\infty} \overline{\bar{\beta}V(x)\alpha} dx = \int_{-\infty}^{\infty} \beta V(x)\bar{\alpha} dx, \tag{2.25}$$

and therefore

$$I_1 + I_2 = \int_{-\infty}^{\infty} \beta H\bar{\alpha} dx. \tag{2.26}$$

6. We have:

$$\langle x \rangle = \int_{-\infty}^{\infty} x\bar{\psi}\psi dx, \tag{2.27}$$

and

$$\langle p \rangle = m\frac{d\langle x \rangle}{dt} = m\int_{-\infty}^{\infty} x\left( \frac{\partial\bar{\psi}}{\partial t}\psi + \bar{\psi}\frac{\partial\psi}{\partial t} \right) dx. \tag{2.28}$$

We use the time dependent Schrödinger equation to eliminate the time derivatives in the integrand in the usal manner:

$$\frac{\partial\psi}{\partial t} = \frac{1}{i\hbar}\left( -\frac{\hbar^2}{2m}\frac{\partial^2\psi}{\partial x^2} + V(x)\psi \right) \tag{2.29}$$

and

$$\frac{\partial\bar{\psi}}{\partial t} = -\frac{1}{i\hbar}\left( -\frac{\hbar^2}{2m}\frac{\partial^2\bar{\psi}}{\partial x^2} + V(x)\bar{\psi} \right) \tag{2.30}$$

and therefore the integrand for $\langle p \rangle$ becomes:

$$x\left( \frac{\partial\bar{\psi}}{\partial t}\psi + \bar{\psi}\frac{\partial\psi}{\partial t} \right) = \frac{\hbar}{2mi} x\left( \psi\frac{\partial^2\bar{\psi}}{\partial x^2} - \bar{\psi}\frac{\partial^2\psi}{\partial x^2} \right). \tag{2.31}$$

Now we use the expression obtained for $I_1$ derived in the previous problem to obtain:

$$\langle p \rangle = m \int_{-\infty}^{\infty} x \left( \frac{\partial \bar{\psi}}{\partial t} \psi + \bar{\psi} \frac{\partial \psi}{\partial t} \right) dx,$$

$$= \frac{\hbar}{2i} \int_{-\infty}^{\infty} x\psi \frac{\partial^2 \bar{\psi}}{\partial x^2} - x\bar{\psi} \frac{\partial^2 \psi}{\partial x^2} dx,$$

$$= \frac{\hbar}{2i} \int_{-\infty}^{\infty} \bar{\psi} \frac{\partial^2}{\partial x^2} (x\psi) - x\bar{\psi} \frac{\partial^2 \psi}{\partial x^2} dx,$$

we used the expression for $I_1$ from the previous problem

$$= -i\hbar \int_{-\infty}^{\infty} \bar{\psi} \frac{\partial \psi}{\partial x} dx. \tag{2.32}$$

7. We have

$$i\hbar \frac{d}{dt} \int_{-\infty}^{\infty} \bar{\psi} A \psi dx = i\hbar \int_{-\infty}^{\infty} \frac{\partial \bar{\psi}}{\partial t} A\psi dx + i\hbar \int_{-\infty}^{\infty} \bar{\psi} A \frac{\partial \psi}{\partial t} dx. \tag{2.33}$$

Using the Schrödinger equation we have:

$$i\hbar \frac{\partial \psi}{\partial t} = H\psi, \tag{2.34}$$

and

$$-i\hbar \frac{\partial \bar{\psi}}{\partial t} = \overline{(H\psi)}. \tag{2.35}$$

Therefore

$$i\hbar \frac{d}{dt} \int_{-\infty}^{\infty} \bar{\psi} A \psi dx = -\int_{-\infty}^{\infty} \overline{H\psi} A\psi dx + \int_{-\infty}^{\infty} \bar{\psi} A H \psi dx,$$

$$= -\int_{-\infty}^{\infty} \bar{\psi} H A\psi dx + \int_{-\infty}^{\infty} \bar{\psi} A H \psi dx,$$

using Hermiticity,

$$= \int_{-\infty}^{\infty} \bar{\psi} (AH - HA) \psi dx = \int_{-\infty}^{\infty} \bar{\psi} [A, H] \psi dx. \tag{2.36}$$

1. The expectation value of the energy is the expectation value of the Hamiltonian operator, which has the form:

$$\langle E \rangle = \int_{-\infty}^{\infty} \bar{\psi} H \psi dx. \tag{2.37}$$

If $\frac{\partial V}{\partial t} = 0$ then $\frac{\partial H}{\partial t} = 0$, so the Hamiltonian does not depend explicitly on time. Hence we can set $A = H$ and apply the result above to obtain:

$$\frac{d}{dt}\langle E \rangle = \frac{1}{i\hbar}\int_{-\infty}^{\infty} \bar{\psi}[H,H]\psi\, dx = 0 \tag{2.38}$$

since $[H,H] = 0$. Therefore $\langle E \rangle = $ constant. This result corresponds to conservation of energy (note that $V$ is not explicitly time dependent).

2. The time derivative of the expectation value for $p$ has the form:

$$\frac{d}{dt}\langle p \rangle = -i\hbar\frac{d}{dt}\int_{-\infty}^{\infty} \bar{\psi}\frac{\partial}{\partial x}\psi\, dx. \tag{2.39}$$

We can apply the general result obtained at the beginning of this problem by taking $A = \frac{\partial}{\partial x}$. Hence,

$$\begin{aligned}
[A,H]\psi &= \left[\frac{\partial}{\partial x}\left(-\frac{\hbar^2}{2m}\frac{\partial^2}{\partial x^2} + V(x)\right)\right.\\
&\quad \left. -\left(-\frac{\hbar^2}{2m}\frac{\partial^2}{\partial x^2} + V(x)\right)\frac{\partial}{\partial x}\right]\psi,\\
&= \frac{\partial}{\partial x}(V\psi) - V\frac{\partial\psi}{\partial x} = \frac{\partial V}{\partial x}\psi, \tag{2.40}
\end{aligned}$$

and therefore

$$\frac{d}{dt}\langle p \rangle = -\int_{-\infty}^{\infty}\bar{\psi}\frac{dV}{dx}\psi\, dx = -\left\langle\frac{\partial V}{\partial x}\right\rangle. \tag{2.41}$$

This relation is analogous to Newton's second law.

8. In problem 1 we derived:

$$\psi(x,t) = \frac{1}{\sqrt{2}}\psi_1(x)\exp\left(-\frac{iE_1 t}{\hbar}\right) + \frac{1}{\sqrt{2}}\psi_2(x)\exp\left(-\frac{iE_2 t}{\hbar}\right). \tag{2.42}$$

From which we obtain:

$$i\hbar\frac{\partial\psi}{\partial t} = \frac{1}{\sqrt{2}}E_1\psi_1(x)\exp\left(-\frac{iE_1 t}{\hbar}\right) + \frac{1}{\sqrt{2}}E_2\psi_2(x)\exp\left(-\frac{iE_2 t}{\hbar}\right). \tag{2.43}$$

Using

$$\int_0^a \psi_n(x)\psi_m(x)\, dx = \delta_{nm}, \tag{2.44}$$

we obtain

$$i\hbar \int_0^a \bar{\psi}\frac{\partial\psi}{\partial t}dx = \frac{1}{2}(E_1 + E_2) = \frac{\pi^2\hbar^2}{4ma^2}(1+4) = \frac{5\pi^2\hbar^2}{4ma^2} \tag{2.45}$$

where $\frac{1}{2}$ is the probability to be in the first eigenstate and $\frac{1}{2}$ is the probability to be in the second eigenstate. Therefore the result is equivalent to the sum of $E_1$ times the probability to be in the first eigenstate and $E_2$ times the probability to be in the second eigenstate.

9. Classically, all particles incident from $-\infty$ continue to $x > 0$, i.e. there is no reflection. Solutions of the time independent Schrödinger equation are given by:

$$x < 0 \quad \psi(x) = A\exp(ik_1x) + B\exp(-ik_1x),$$
$$x > 0 \quad \psi(x) = C\exp(ik_2x), \tag{2.46}$$

where

$$k_1 = \frac{1}{\hbar}\sqrt{2mE}, \quad k_2 = \frac{1}{\hbar}\sqrt{2m(E+V)}. \tag{2.47}$$

We apply the matching conditions at $x = 0$:

$$\text{continuity of } \psi \quad \text{at} \quad x = 0 \quad \rightarrow \quad A + B = C,$$
$$\text{continuity of } \psi' \quad \text{at} \quad x = 0 \quad \rightarrow \quad k_1(A - B) = k_2C, \tag{2.48}$$

From these relations we obtain:

$$\frac{B}{A} = \frac{1 - \frac{k_2}{k_1}}{1 + \frac{k_2}{k_1}}. \tag{2.49}$$

and therefore

$$R = \left|\frac{B}{A}\right|^2 = \left(\frac{1 - \frac{k_2}{k_1}}{1 + \frac{k_2}{k_1}}\right)^2. \tag{2.50}$$

It then follows that

$$T = 1 - R = \frac{4\frac{k_2}{k_1}}{\left(1 + \frac{k_2}{k_1}\right)^2}. \tag{2.51}$$

Using (2.47), we have:

$$\frac{k_2}{k_1} = \frac{\sqrt{E+V}}{\sqrt{E}} = \sqrt{1 + \frac{V}{E}}$$

Therefore $R$ can be rewritten as:

$$R = \left( \frac{1 - \sqrt{1 + \frac{V}{E}}}{1 + \sqrt{1 + \frac{V}{E}}} \right)^2 = \left( \frac{\frac{1}{\sqrt{1 + \frac{V}{E}}} - 1}{\frac{1}{\sqrt{1 + \frac{V}{E}}} + 1} \right)^2 , \tag{2.52}$$

from which it follows that:

$$\lim_{\frac{V}{E} \to \infty} R = 1. \tag{2.53}$$

This is a purely quantum mechanical result.

10. Solutions of the time independent Schrödinger equation are given by

$$\psi(x) = \begin{cases} A \exp(ikx) + B \exp(-ikx), & x < 0, \\ C \exp(ik_1 x) + D \exp(-ik_1 x), & 0 < x < a, \\ E \exp(ikx), & x > a, \end{cases} \tag{2.54}$$

where

$$k = \frac{1}{\hbar} \sqrt{2mE}, \quad k_1 = \frac{1}{\hbar} \sqrt{2m(E - V)}. \tag{2.55}$$

Continuity of $\psi(x)$ and $\psi'(x)$ implies:

$$A + B = C + D,$$

$$k(A - B) = k_1(C - D),$$

$$C \exp(ik_1 a) + D \exp(-ik_1 a) = E \exp(ika),$$

$$k_1(C \exp(ik_1 a) - D \exp(-ik_1 a)) = Ek \exp(ika) \tag{2.56}$$

Solving for $\frac{A}{B}$ we find

$$R = \left| \frac{B}{A} \right|^2 = \frac{1}{1 + \frac{4k^2 k_1^2}{(k^2 - k_1^2)^2 \sin^2(k_1 a)}}, \tag{2.57}$$

from which we can obtain:

$$T = 1 - R = \frac{1}{\frac{(k^2 - k_1^2)^2 \sin^2(k_1 a)}{4k^2 k_1^2}}. \tag{2.58}$$

Now we notice that $T = 1$ when $\sin k_1 a = 0$, i.e., $k_1 a = n\pi$, which implies $\sqrt{2m(E - V)} = n\pi\hbar$.

11. (a)

$$|\Psi|^2 = \Psi\overline{\Psi} = |A|^2(\psi_1 + \psi_2)(\overline{\psi_1} + \overline{\psi_2})$$
$$= |A|^2(|\psi_1|^2 + \overline{\psi_1}\psi_2 + \psi_1\overline{\psi_2} + |\psi_2|^2)$$

To normalize $\Psi$ we need

$$1 = \int |\Psi|^2 dx = |A|^2 \int (|\psi_1|^2 + \overline{\psi_1}\psi_2 + \psi_1\overline{\psi_2} + |\psi_2|^2)$$
$$= 2|A|^2,$$

where the last equality follows by orthonormality of $\psi_1$ and $\psi_2$. Solving the resulting equation, we get that

$$A = \frac{1}{\sqrt{2}}$$

(b)

$$\Psi(x,t) = \frac{1}{\sqrt{2}}\left(\psi_1 e^{-iE_1 t/\hbar} + \psi_2 e^{-iE_2 t/\hbar}\right)$$
$$= \frac{1}{\sqrt{2}}\left(\sin(\pi x)e^{-i\omega t} + \sin(2\pi x)e^{-i4\omega t}\right)$$
$$|\Psi(x,t)|^2 = \sin^2(\pi x) + \sin(\pi x)\sin(2\pi x)$$
$$\times \left(e^{-3i\omega t} + e^{3i\omega t}\right) + \sin^2(2\pi x)$$
$$= \sin^2(\pi x) + \sin^2(2\pi x) + 2\sin(\pi x)$$
$$\times \sin(2\pi x)\cos(3\omega t)$$

(c) Taking the expectation of $x$ we get

$$\langle x \rangle = \int x|\Psi(x,t)|^2 dx = \int_0^1 x\left(\sin^2(\pi x) + \sin^2(2\pi x)\right.$$
$$\left. + 2\sin(\pi x)\sin(2\pi x)\cos(3\omega t)\right) dx$$

Now using trigonometric identities in the hint, we get that

$$\int_0^1 x\sin^2(\pi x)dx = \left(\frac{x^2}{4} - \frac{x\sin(2\pi x)}{4\pi} - \frac{\cos(2\pi x)}{8\pi^2}\right)\Bigg|_0^1$$
$$= \frac{1}{4} = \int_0^1 x\sin^2(2\pi x)dx$$

and

$$\int_0^1 x\sin(\pi x)\sin(2\pi x) = \frac{1}{2}\int_0^1 x(\cos(\pi x) - \cos(3\pi x))dx$$

$$= \frac{1}{2}\left(\frac{1}{\pi^2}\cos(\pi x) + \frac{x}{\pi}\sin(\pi x)\right.$$

$$\left.\left. - \frac{1}{9\pi^2}\cos(3\pi x) - \frac{x}{3\pi}\sin(3\pi x)\right)\right|_0^1$$

$$= -\frac{8}{9\pi^2}$$

Putting it all together, we get that

$$\langle x \rangle = \frac{1}{2}\left(1 - \frac{32}{9\pi^2}\cos(3\omega t)\right).$$

(d)

$$\langle p \rangle = m\frac{d\langle x \rangle}{dt} = \frac{m}{2}\frac{32}{9\pi^2}(-3\omega)\sin(3\omega t) = \frac{8\hbar}{3a}\sin(3\omega t)$$

(e) The two possible energies are $E_1 = \pi^2\hbar^2/2ma^2$ or $E_2 = 2\pi^2\hbar^2/ma^2$ with equal probability $P_1 = P_2 = 1/2$. Therefore

$$\langle H \rangle = \frac{1}{2}(E_1 + E_2) = \frac{5\pi^2\hbar^2}{4ma^2}.$$

12. (a)

$$\Psi(x,0)\overline{\Psi(x,0)} = |C|^2\left(|\psi_1|^2 + \psi_2\bar\psi_1 + \psi_1\bar\psi_2 + |\psi_2|^2\right).$$

Integrating this expression from $-\infty$ to $\infty$ and using orthonormality of $\psi_1$ and $\psi_2$ gives:

$$1 = 2|C|^2,$$

or

$$C = \frac{1}{\sqrt{2}}.$$

(b)

$$\Psi(x,t) = \frac{1}{\sqrt{2}}\left(\psi_1(x)e^{-\frac{iE_1 t}{\hbar}} + \psi_2(x)e^{-\frac{iE_2 t}{\hbar}}\right).$$

(c)

$$\langle E \rangle = i\hbar \int_{-\infty}^{\infty} \overline{\Psi(x,t)} \frac{\partial}{\partial t} \Psi(x,t)\, dx,$$

The integrand of this integral has the form:

$$-\frac{1}{2}\left(\psi_1(x)e^{\frac{iE_1 t}{\hbar}} + \psi_2(x)e^{\frac{iE_2 t}{\hbar}}\right)$$

$$\times \left(\frac{iE_1}{\hbar}\psi_1(x)e^{-\frac{iE_1 t}{\hbar}} + \frac{iE_2}{\hbar}\psi_2(x)e^{-\frac{iE_2 t}{\hbar}}\right),$$

$$= -\frac{1}{2}\left(\frac{iE_1}{\hbar}\psi_1^2 + \frac{iE_2}{\hbar}\psi_2^2 + \frac{iE_2}{\hbar}\psi_1\psi_2 e^{\frac{i(E_1-E_2)t}{\hbar}}\right.$$

$$\left. + \frac{iE_1}{\hbar}\psi_1\psi_2 e^{\frac{-i(E_1-E_2)t}{\hbar}}\right)$$

Integrating this expression, and using orthonormality of $\psi_1$ and $\psi_2$, and the fact that they are real, gives:

$$\langle E \rangle = \frac{1}{2}(E_1 + E_2).$$

(d)

$$\langle E^2 \rangle = -\hbar^2 \int_{-\infty}^{\infty} \overline{\Psi(x,t)} \frac{\partial^2}{\partial t^2} \Psi(x,t)\, dx.$$

The integrand of this integral has the form:

$$\overline{\Psi(x,t)}\frac{\partial^2}{\partial t^2}\Psi(x,t) = -\frac{1}{2}\left(\psi_1(x)e^{\frac{iE_1 t}{\hbar}} + \psi_2(x)e^{\frac{iE_2 t}{\hbar}}\right)$$

$$\times \left(\frac{E_1^2}{\hbar^2}\psi_1(x)e^{\frac{-iE_1 t}{\hbar}} + \frac{E_2^2}{\hbar^2}\psi_2(x)e^{\frac{-iE_2 t}{\hbar}}\right).$$

Computing the integral gives:

$$\langle E^2 \rangle = \frac{\hbar^2}{2}\left(\frac{E_1^2}{\hbar^2} + \frac{E_2^2}{\hbar^2}\right) = \frac{1}{2}\left(E_1^2 + E_2^2\right).$$

Therefore we have:

$$(\Delta E)^2 = \langle E^2 \rangle - \langle E \rangle^2 = \frac{1}{2}\left(E_1^2 + E_2^2\right) - \frac{1}{4}(E_1 + E_2)^2$$

$$= \frac{1}{4}(E_2 - E_1)^2,$$

and therefore

$$\Delta E = \frac{1}{2}(E_2 - E_1).$$

(e) We have:

$$\langle x \rangle = \int_{-\infty}^{\infty} x |\Psi(x,t)|^2 dx,$$

$$= \frac{1}{2} \int_{-\infty}^{\infty} x \left( |\psi_1|^2 + |\psi_2|^2 + 2\psi_1\psi_2 \cos\left(\frac{(E_1 - E_2)t}{\hbar}\right) \right) dx.$$

We identify the following constant terms in the integral:

$$x_0 = \frac{1}{2} \int_{-\infty}^{\infty} x \left( |\psi_1|^2 + |\psi_2|^2 \right) dx,$$

and

$$a = \int_{-\infty}^{\infty} x\psi_1\psi_2 dx.$$

Therefore we have:

$$\langle x \rangle = x_0 + a \cos\left(\frac{(E_1 - E_2)t}{\hbar}\right).$$

13. (a) We argue by contradiction. Let $\psi_n$ be an eigenstate with eigenvalue $E_n$ and $\psi_m$ be a *different* eigenstate with eigenvalue $E_m$, and we assume that $E_n = E_m = E$. In this case we have

$$-\frac{\hbar^2}{2m}\frac{d^2\psi_n}{dx^2} + V(x)\psi_n = E\psi_n,$$

$$-\frac{\hbar^2}{2m}\frac{d^2\psi_m}{dx^2} + V(x)\psi_m = E\psi_m. \tag{2.59}$$

Multiplying the first equation by $\psi_m$, the second by $\psi_n$ and subtracting gives:

$$\psi_m\psi_n'' - \psi_n\psi_m'' = 0, \tag{2.60}$$

which can be rewritten as:

$$\frac{d}{dx}\left(\psi_m\psi_n' - \psi_n\psi_m'\right) = 0, \tag{2.61}$$

from which it follows that:

$$\psi_m\psi_n' - \psi_n\psi_m' = \text{constant}. \tag{2.62}$$

Now since $\psi_n \to 0$, $\psi_m \to 0$ as $|x| \to \infty$ it follows that constant $= 0$, which further implies that:

$$\psi_m \psi_n' - \psi_n \psi_m' = 0, \tag{2.63}$$

or

$$\frac{d}{dx} \log \psi_n = \frac{d}{dx} \log \psi_m. \tag{2.64}$$

Integrating this expression gives:

$$\psi_n = \text{constant } \psi_m. \tag{2.65}$$

This implies that $\psi_n$ and $\psi_m$ are the same state. But this a contradiction that comes from the assumption that $E_n = E_m$.

(b) Suppose $\psi$ satisifies the one dimensional time independent Schrödinger equation with eigenvalue $E$. By taking the complex conjugate of the one dimensional time independent Schrödinger equation it is easy to see that $\overline{\psi}$ is also a solution with the same eigenvalue $E$. Hence if $\psi$ and $\overline{\psi}$ are different, then we have a degeneracy, which would contradict the previous result.

By superposition, the following are two real degenerate solutions.

$$\psi_r = \frac{1}{2}\left(\psi + \overline{\psi}\right), \qquad \psi_{im} = \frac{1}{2i}\left(\psi - \overline{\psi}\right).$$

Hence by the previous result we have:

$$\psi_{im} = c\psi_r.$$

where $c$ can be taken to be real since $\psi_r$ and $\psi_{im}$ are real. It then follows that

$$\psi = \psi_r + i\psi_{im} = (1 + ic)\psi_r.$$

Writing $1 + ic = \sqrt{1 + c^2}\, e^{i\beta}$, with $\beta$ real, shows that, up to a phase factor, $\psi$ can be chosen to be real.

(c) We have

$$\psi = \psi_r + i\psi_{im},$$

$$\overline{\psi} = \psi_r - i\psi_{im},$$

from which it follows that

$$\overline{\psi}\psi' = \left(\psi_r - i\psi_{im}\right)\left(\psi_r' + i\psi_{im}'\right),$$

$$= \psi_r \psi_r' + \psi_{im}\psi_{im}' + i\left(\psi_r \psi_{im}' - \psi_{im}\psi_r'\right),$$

and

$$\psi\overline{\psi}' = (\psi_r + i\psi_{im})\,(\psi'_r - i\psi'_{im}),$$

$$= \psi_r\psi'_r + \psi_{im}\psi'_{im} + i\,(\psi_{im}\psi'_r - \psi_r\psi'_{im}.),$$

and therefore

$$\frac{\hbar}{2mi}\left(\overline{\psi}\psi' - \psi\overline{\psi}'\right) = \frac{\hbar}{m}\,(\psi_r\psi'_{im} - \psi_{im}\psi'_r). \qquad (2.66)$$

Next we compute:

$$\frac{1}{m}\mathrm{Re}\left(\overline{\psi}\frac{\hbar}{i}\psi'\right) = \frac{1}{m}\mathrm{Re}\,(\psi_r - i\psi_{im})\left(\frac{\hbar}{i}(\psi'_r + i\psi'_{im}\right),$$

$$= \frac{\hbar}{m}\mathrm{Re}\,(\psi_r - i\psi_{im})\,(-i\psi'_r + \psi'_{im}),$$

$$= \frac{\hbar}{m}\mathrm{Re}\,(-i\psi_r\psi'_r + \psi_r\psi'_{im} - \psi_{im}\psi'_r - i\psi_{im}\psi'_{im}),$$

$$= \frac{\hbar}{m}\,(\psi_r\psi'_{im} - \psi_{im}\psi'_r).$$

It now follows from (2.66) and (2.67) that the result holds.

## Chapter 3

1. (a)

$$[A, B] = AB - BA$$

$$= -(BA - AB) = -[B, A]$$

(b) Let $A = \alpha C + \beta D$ and $B = \delta E + \gamma F$, for operators $C, D, E, F$ and scalars $\alpha, \beta, \delta, \gamma$.

$$[\alpha C + \beta D, \delta E + \gamma F]$$

$$= (\alpha C + \beta D)(\delta E + \gamma F) - (\delta E + \gamma F)(\alpha C + \beta D)$$

$$= \alpha\delta CE + \alpha\gamma CF + \beta\delta DE + \beta\gamma DF - \delta\alpha EC$$

$$-\delta\beta ED - \gamma\alpha FC - \gamma\beta FD$$

$$= \alpha\delta(CE - EC) + \alpha\gamma(CF - FC) + \beta\delta(DE - ED)$$

$$-\beta\gamma(DF - FD)$$

$$= \alpha\delta[C, E] + \alpha\gamma[C, F] + \beta\delta[D, E] + \beta\gamma[D, f]$$

Therefore the commutator is linear in both the first and second operator.

(c)

$$[A, BC] = ABC - BCA$$

$$= ABC - BAC + BAC - BCA$$

$$= [A, B]C + B[A, C]$$

(d)

$$[A, [B, C]] + [B, [C, A]] + [C, [A, B]]$$

$$= A(BC - CB) - (BC - CB)A + B(CA - AC)$$

$$-(CA - AC)B + C(AB - BA) - (AB - BA)C$$

$$= ABC - ACB - BCA + CBA + BCA - BAC - CAB$$

$$+ACB + CAB - CBA - ABC + BAC = 0$$

2. (a) If $n = 1$, $[X, P] = i\hbar$. Assume that $[X^n, P] = i\hbar n X^{n-1}$. Then

$$[X^{n+1}, P] = X^{n+1}P - PX^{n+1}$$

$$= X^{n+1}P - (XP - i\hbar)X^n$$

$$= X(X^n P - PX^n) + i\hbar X^n$$

$$= X[X^n, P] + i\hbar X^n$$

$$= X i\hbar n X^{n-1} + i\hbar X^n$$

$$= i\hbar(n + 1)X^n.$$

Since this relation holds for $n = 1$, it is proven by induction to be true for all positive integers $n$.

Similarly, assume that $[X, P^n] = i\hbar n P^{n-1}$, then

$$[X, P^{n+1}] = XP^{n+1} - P^{n+1}X$$

$$= (PX + i\hbar)P^n - P^{n+1}X$$

$$= P(XP^n - P^n X) + i\hbar P^n$$

$$= P[X, P^n] + i\hbar P^n$$

$$= P i\hbar n P^{n-1} + i\hbar P^n$$

$$= i\hbar(n + 1)P^n$$

Since this relation holds for $n = 1$, it is proven by induction to be true for all positive integers $n$.

(b) Let $f(X) = a_0 + a_1 X + a_2 X^2 + \cdots + a_m X^m$. Then

$$[f(X), P] = a_0[1, P] + a_1[X, P] + a_2[X^2, P]$$
$$+ \cdots + a_m[X^m, P]$$
$$= 0 + a_1 i\hbar + 2i\hbar a_2 X + \cdots + m i\hbar a_m X^{m-1}$$
$$= i\hbar(a_1 + 2a_2 X + 3a_3 X^2 + \cdots + m a_m X^{m-1})$$
$$= i\hbar f'(X)$$

Let $g(P) = b_0 + b_1 P + b_2 P^2 + \cdots + b_k P^k$. Then

$$[X, g(P)] = b_0[X, 1] + b_1[X, P] + b_2[X, P^2]$$
$$+ \cdots + b_k[X, P^k]$$
$$= 0 + b_1 i\hbar + b_2 2i\hbar P + \cdots + b_k k i\hbar x^{k-1}$$
$$= i\hbar(b_1 + 2b_2 P + 3b_3 P^2 + \cdots + k b_k P^{k-1})$$
$$= i\hbar g'(P).$$

(c) Start with $n = 2$.

$$[A, B^2] = B[A, B] + [A, B]B$$

by part (c) of problem 1 above. But $B$ commutes with $[A, B]$, so $[A, B]B = B[A, B]$. Therefore

$$[A, B^2] = 2B[A, B].$$

If it is true that $[A, B^n] = nB^{n-1}[A, B]$, then

$$[A, B^{n+1}] = B^n[A, B] + [A, B^n]B$$
$$= B^n[A, B] + (nB^{n-1}[A, B])B$$
$$= (n + 1)B^n[A, B].$$

Thus by induction we have proven that $[A, B^n] = nB^{n-1}[A, B]$. Exchanging the place of $A$ and $B$, it follows in exactly the same way that

$$[B, A^n] = nA^{n-1}[B, A]$$
$$= -nA^{n-1}[A, B].$$

3. (a) We start with

$$\frac{d}{d\theta}(e^{\theta N}) = \frac{d}{d\theta}\sum_{n=0}^{\infty}\frac{(\theta N)^n}{n!}.$$

$\theta$ is just a parameter (not an operator) so $(\theta N)^n = \theta^n N^n$. Therefore,

$$\begin{aligned}
\frac{d}{d\theta}(e^{\theta N}) &= \sum_{n=0}^{\infty}\frac{(\frac{d}{d\theta}\theta^n)N^n}{n!} \\
&= \sum_{n=1}^{\infty}\frac{n\theta^{n-1}N^n}{n!} \\
&= N\sum_{n=1}^{\infty}\frac{\theta^{n-1}N^{n-1}}{(n-1)!} \\
&= N\sum_{n=0}^{\infty}\frac{\theta^n N^n}{n!} = Ne^{\theta N}.
\end{aligned}$$

(b)

$$\begin{aligned}
\frac{d}{d\theta}(e^{\theta N}Ae^{-\theta N}) &= Ne^{\theta N}Ae^{-\theta N} + e^{\theta N}A(-N)e^{-\theta N} \\
&= e^{\theta N}(NA - AN)e^{-\theta N} \\
&= -e^{\theta N}Ae^{-\theta N} \\
&= -C(\theta)
\end{aligned}$$

(c) Let $T(\theta) = e^{-\theta}A$. We check that it is a solution of the differential equation:

$$\frac{dT(\theta)}{d\theta} = -e^{-\theta}A = -T(\theta).$$

It also satisfies

$$T(0) = A = C(0).$$

If the solution with this initial condition is unique, then $T(\theta) = C(\theta)$, meaning that

$$C(\theta) = e^{-\theta}A.$$

4. (a)

$$[a, a^\dagger] = [P - im\omega X, P + im\omega X]$$
$$= [P, P] + im\omega[P, X] - im\omega[X, P] + m^2\omega^2[X, X]$$
$$= im\omega(-i\hbar) - im\omega(i\hbar) = 2m\omega\hbar.$$

(b) If $n = 1$ then $aa^\dagger = a^\dagger a + 2m\omega\hbar$.
If $n = 2$, then

$$a(a^\dagger)^2 = (a^\dagger a + 2m\omega\hbar)a^\dagger$$
$$= a^\dagger(a^\dagger a + 2m\omega\hbar) + 2m\omega\hbar a^\dagger$$
$$= (a^\dagger)^2 a + 4m\omega\hbar a^\dagger.$$

So, by induction, if we assume that $a(a^\dagger)^n = (a^\dagger)^n a + n2m\omega\hbar(a^\dagger)^{n-1}$ then

$$a(a^\dagger)^{n+1} = (a^\dagger)^n aa^\dagger + n2m\omega\hbar(a^\dagger)^n$$
$$= (a^\dagger)^n(a^\dagger a + 2m\omega\hbar) + n2m\omega\hbar(a^\dagger)^n$$
$$= (a^\dagger)^{n+1}a + (n+1)2m\omega\hbar(a^\dagger)^n.$$

This proves the statement by induction.

(c)

$$ae^{\lambda a^\dagger} = a\sum_{n=0}^{\infty}\frac{\lambda^n(a^\dagger)^n}{n!}$$
$$= \sum_{n=0}^{\infty}\frac{\lambda^n a(a^\dagger)^n}{n!}$$
$$= \sum_{n=0}^{\infty}\frac{\lambda^n((a^\dagger)^n a + n2m\omega\hbar(a^\dagger)^{n-1})}{n!}$$
$$= \left(\sum_{n=0}^{\infty}\frac{\lambda^n(a^\dagger)^n}{n!}\right)a + 2m\omega\hbar\lambda\sum_{n=1}^{\infty}\frac{\lambda^{n-1}(a^\dagger)^{n-1}}{(n-1)!}$$
$$= e^{\lambda a^\dagger}(a + 2m\omega\hbar\lambda).$$

5. (a) Look at the norm squared of the vector $|n\rangle$:

$$\langle n|n\rangle = |c_n|^2 \langle 0|a^n (a^\dagger)^n|0\rangle$$

$$= |c_n|^2 \langle 0|a^{n-1}\left((a^\dagger)^n a + n2m\omega\hbar(a^\dagger)^{n-1}\right)|0\rangle$$

$$= |c_n|^2(\langle 0|a^{n-1}(a^\dagger)^n a|0\rangle + n2m\omega\hbar$$

$$\langle 0|a^{n-1}(a^\dagger)^{n-1}|0\rangle.$$

But we can choose $c_n$ to be real, and $a|0\rangle = 0$, so the first term is zero.

$$\langle n|n\rangle = c_n^2 n2m\omega\hbar\langle 0|a^{n-1}(a^\dagger)^{n-1}|0\rangle.$$

But

$$|n-1\rangle = c_{n-1}(a^\dagger)^{n-1}|0\rangle,$$

which means that

$$\langle 0|a^{n-1}(a^\dagger)^{n-1}|0\rangle = \frac{1}{c_{n-1}^2}\langle n-1|n-1\rangle.$$

Thus

$$\langle n|n\rangle = \frac{c_n^2}{c_{n-1}^2}n2m\omega\hbar\langle n-1|n-1\rangle,$$

but $|n\rangle$ and $|n-1\rangle$ are normalised so

$$c_{n-1}^2 = 2m\omega\hbar n c_n^2.$$

When $n=0$, $c_0 = 1$. Thus $c_1^2 = 1/(2m\omega\hbar)$ or $c_1 = 1/\sqrt{2m\omega\hbar}$. Similarly,

$$c_2^2 = \frac{1}{2m\omega\hbar}\frac{1}{2m\omega\hbar \cdot 2},$$

and so on, up to

$$c_n = \frac{1}{\sqrt{n!}(\sqrt{2m\omega\hbar})^n}.$$

(b)

$$a|n\rangle = c_n a(a^\dagger)^n|0\rangle$$

$$= c_n((a^\dagger)^n a + n2m\omega\hbar(a^\dagger)^{n-1})|0\rangle$$

$$= c_n n2m\omega\hbar(a^\dagger)^{n-1}|0\rangle$$

$$= \frac{c_n}{c_{n-1}}n2m\omega\hbar|n-1\rangle$$

$$= \frac{\sqrt{(n-1)!}(\sqrt{2m\omega\hbar})^{n-1}}{\sqrt{n!}(\sqrt{2m\omega\hbar})^n}n2m\omega\hbar|n-1\rangle$$

$$= \sqrt{2m\omega\hbar n}|n-1\rangle.$$

(c) We have

$$\langle n_1 | n_2 \rangle = c_{n_1} c_{n_2} \langle 0 | a^{n_1} (a^\dagger)^{n_2} | 0 \rangle.$$

Let $n_2 < n_1$. Then

$$
\begin{aligned}
&\langle 0 | a^{n_1} (a^\dagger)^{n_2} | 0 \rangle \\
&= \langle 0 | a^{n_1 - 1} ((a^\dagger)^{n_2} a + n_2 2 m \omega \hbar (a^\dagger)^{n_2 - 1}) | 0 \rangle \\
&= n_2 2 m \omega \hbar \langle 0 | a^{n_1 - 1} (a^\dagger)^{n_2 - 1} | 0 \rangle.
\end{aligned}
$$

We repeat this process until the exponent on $a^\dagger$ is zero and on $a$ is $n_1 - n_2 > 0$. But $a$ acting on $|0\rangle$ is zero, so $\langle n_1 | n_2 \rangle = 0$ if $n_2 < n_1$. If $n_1 < n_2$ then

$$\langle 0 | a^{n_1} (a^\dagger)^{n_2} | 0 \rangle = \overline{\langle 0 | a^{n_2} (a^\dagger)^{n_1} | 0 \rangle},$$

but the bracket is zero from the calculation above, so its complex conjugate is also zero.

So $\langle n_1 | n_2 \rangle = 0$ if $n_1 \neq n_2$.

(d)

$$
\begin{aligned}
\langle n_1 | X | n_2 \rangle &= \langle n_1 | \frac{a^\dagger - a}{2 i m \omega} | n_2 \rangle \\
&= \frac{1}{2 i m \omega} (\langle n_1 | a^\dagger | n_2 \rangle - \langle n_1 | a | n_2 \rangle) \\
&= \frac{1}{2 i m \omega} \left( \sqrt{(n_2 + 1) 2 m \omega \hbar} \langle n_1 | n_2 + 1 \rangle \right. \\
&\qquad\qquad \left. - \sqrt{2 m \omega \hbar n_2} \langle n_1 | n_2 - 1 \rangle \right) \\
&= \begin{cases} \frac{1}{2 i m \omega} \sqrt{(n_2 + 1) 2 m \omega \hbar} & \text{if } n_2 + 1 = n_1 \\ -\frac{1}{2 i m \omega} \sqrt{2 m \omega \hbar n_2} & \text{if } n_2 - 1 = n_1 \\ 0 & \text{otherwise} \end{cases}
\end{aligned}
$$

$$\langle n_1|P|n_2\rangle = \langle n_1|\frac{a^\dagger + a}{2}|n_2\rangle$$

$$= \frac{1}{2}(\langle n_1|a^\dagger|n_2\rangle + \langle n_1|a|n_2\rangle)$$

$$= \frac{1}{2}\left(\sqrt{(n_2+1)2m\omega\hbar}\langle n_1|n_2+1\rangle\right.$$

$$\left. + \sqrt{2m\omega\hbar n_2}\langle n_1|n_2-1\rangle\right)$$

$$= \begin{cases} \frac{1}{2}\sqrt{(n_2+1)2m\omega\hbar} & \text{if } n_2+1=n_1 \\ \frac{1}{2}\sqrt{2m\omega\hbar n_2} & \text{if } n_1=n_2-1 \\ 0 & \text{otherwise} \end{cases}$$

6. If $|\psi\rangle$ is an eigenstate, the eigenvalue is $E_{|\psi\rangle}(A)$, the expectation value of $A$ in the state $|\psi\rangle$. So the dispersion is

$$\Delta_{|\psi\rangle}(A) = [E_{|\psi\rangle}(A^2) - E_{|\psi\rangle}(A)^2]^{1/2} \tag{3.1}$$

$$= \left[\frac{\langle\psi|A(A\psi)\rangle}{||\,|\psi\rangle\,||^2} - (E_{|\psi\rangle}(A))^2\right]^{1/2} \tag{3.2}$$

$$= \left[\frac{E_{|\psi\rangle}(A)\langle\psi|A\psi\rangle}{||\,|\psi\rangle\,||^2} - (E_{|\psi\rangle}(A))^2\right]^{1/2} \tag{3.3}$$

$$= [(E_{|\psi\rangle}(A))^2 - (E_{|\psi\rangle}(A))^2]^{1/2} = 0 \tag{3.4}$$

On the other hand, if the dispersion is 0

$$0 = \Delta_{|\psi\rangle}(A)^2 \tag{3.5}$$

$$= E_{|\psi\rangle}((A - E_{|\psi\rangle}(A))^2) \tag{3.6}$$

$$= \frac{\langle\psi|(A - E_\psi(A))^2|\psi\rangle}{||\,|\psi\rangle\,||^2} \tag{3.7}$$

$$= \frac{\langle\psi|(A - E_\psi(A))^\dagger(A - E_{|\psi\rangle}(A))|\psi\rangle}{||\,|\psi\rangle\,||^2} \tag{3.8}$$

(because $A$ is self-adjoint)

$$0 = \frac{||(A - E_{|\psi\rangle}(A))\,|\psi\rangle\,||^2}{||\,|\psi\rangle\,||^2} \tag{3.9}$$

$$\Rightarrow 0 = (A - E_{|\psi\rangle}(A))\,|\psi\rangle \tag{3.10}$$

$$\Rightarrow A\,|\psi\rangle = E_{|\psi\rangle}(A)\,|\psi\rangle\,, \tag{3.11}$$

which means that $|\psi\rangle$ is an eigenvector if the dispersion is zero.

7. (a) As a matrix, we write

$$G = \begin{pmatrix} \frac{5}{4} & \frac{\sqrt{3}}{4} & 0 \\ \frac{\sqrt{3}}{4} & \frac{7}{4} & 0 \\ 0 & 0 & 2 \end{pmatrix}.$$

To calculate the eigenvalues, let

$$\det \begin{pmatrix} \frac{5}{4} - \lambda & \frac{\sqrt{3}}{4} & 0 \\ \frac{\sqrt{3}}{4} & \frac{7}{4} - \lambda & 0 \\ 0 & 0 & 2 - \lambda \end{pmatrix} = 0$$

$$(2 - \lambda)[(5/4 - \lambda)(7/4 - \lambda) - 3/16] = 0$$

$$(2 - \lambda)(\lambda^2 - 3\lambda + 2) = 0$$

$$(2 - \lambda)(\lambda - 1)(\lambda - 2) = 0,$$

so $\lambda = 1$ or $\lambda = 2$.

If $\lambda = 1$, then

$$\begin{pmatrix} \frac{1}{4} & \frac{\sqrt{3}}{4} & 0 \\ \frac{\sqrt{3}}{4} & \frac{3}{4} & 0 \\ 0 & 0 & 1 \end{pmatrix} \begin{pmatrix} a \\ b \\ c \end{pmatrix} = \begin{pmatrix} 0 \\ 0 \\ 0 \end{pmatrix}$$

and so $a = -\sqrt{3}b$, $c = 0$. The general eigenvector is $(-\sqrt{3}b, b, 0)$. Normalised, we have

$$|e_1\rangle = -\frac{\sqrt{3}}{2}|1\rangle + \frac{1}{2}|2\rangle.$$

If $\lambda = 2$, then

$$\begin{pmatrix} -\frac{3}{4} & \frac{\sqrt{3}}{4} & 0 \\ \frac{\sqrt{3}}{4} & -\frac{1}{4} & 0 \\ 0 & 0 & 0 \end{pmatrix} \begin{pmatrix} a \\ b \\ c \end{pmatrix} = \begin{pmatrix} 0 \\ 0 \\ 0 \end{pmatrix}$$

and so $a = b/\sqrt{3}$, with no condition on $c$. The general eigenvector is $(b/\sqrt{3}, b, c)$.

We want an orthonomal basis. Note that all eigenvectors of the form $(b/\sqrt{3}, b, c)$ are orthogonal to the eigenvector for $\lambda = 1$, $(-\sqrt{3}/2, 1/2, 0)$:

$$(-\sqrt{3}/2, 1/2, 0) \cdot (b/\sqrt{3}, b, c) = -b/2 + b/2 + 0 = 0.$$

The space spanned by the eigenvectors corresponding to $\lambda = 2$ is two-dimensional, so we need to find two orthonormal eigenvectors in this space. There are lots of ways to choose them and it doesn't make a difference what you choose, but some choices make life easier than others. Let's let $b = 0$, $c = 1$ for one choice, giving $(0, 0, 1)$, and $b = 1$, $c = 0$ for the other, which when normalised gives $(1/2, \sqrt{3}/2, 0)$. You can check that these two are orthonormal. Thus our set of orthonormal eigenvectors is

$$\lambda = 1 \quad |e_1\rangle = -\frac{\sqrt{3}}{2}|1\rangle + \frac{1}{2}|2\rangle$$

$$\lambda = 2 \quad |e_2\rangle = \frac{1}{2}|1\rangle + \frac{\sqrt{3}}{2}|2\rangle$$

$$|e_3\rangle = |3\rangle.$$

(b) We can write the operator $G$ as

$$G = |e_1\rangle\langle e_1| + 2(|e_2\rangle\langle e_2| + |e_3\rangle\langle e_3|).$$

We calculate the probability of getting the outcome 2. Note that you may have chosen different eigenvectors to span the $\lambda = 2$ space, but the answer should be the same in the end. The probability is

$$\langle\phi|(|e_2\rangle\langle e_2| + |e_3\rangle\langle e_3|)|\phi\rangle$$

$$= (\cos\theta\langle 1| + \sin\theta\langle 3|)\left(\frac{1}{2}|1\rangle + \frac{\sqrt{3}}{2}|2\rangle\right)$$

$$\times \left((1/2\langle 1| + \sqrt{3}/2\langle 2|)\right.$$

$$\left. + |3\rangle\langle 3|)(\cos\theta|1\rangle + \sin\theta|3\rangle)\right.$$

$$= (\tfrac{1}{2}\cos\theta)^2 + \sin^2\theta.$$

The state immediately after measurement is

$$\frac{(|e_2\rangle\langle e_2| + |e_3\rangle\langle e_3|)|\phi\rangle}{||(|e_2\rangle\langle e_2| + |e_3\rangle\langle e_3|)|\phi\rangle||}$$

$$= \frac{\begin{array}{c}(\frac{1}{2}|1\rangle + \frac{\sqrt{3}}{2}|2\rangle)(\frac{1}{2}\langle 1| + \frac{\sqrt{3}}{2}\langle 2|) \\ (\cos\theta|1\rangle + \sin\theta|3\rangle) \\ +|3\rangle\langle 3|(\cos\theta|1\rangle + \sin\theta|3\rangle)\end{array}}{\begin{array}{c}||(\frac{1}{2}|1\rangle + \frac{\sqrt{3}}{2}|2\rangle) \\ (\frac{1}{2}\langle 1| + \frac{\sqrt{3}}{2}\langle 2|)(\cos\theta|1\rangle + \sin\theta|3\rangle) \\ +|3\rangle\langle 3|(\cos\theta|1\rangle + \sin\theta|3\rangle)||\end{array}}$$

$$= \frac{\frac{1}{4}\cos\theta|1\rangle + \frac{\sqrt{3}}{4}\cos\theta|2\rangle + \sin\theta|3\rangle}{||\frac{1}{4}\cos\theta|1\rangle + \frac{\sqrt{3}}{4}\cos\theta|2\rangle + \sin\theta|3\rangle||}$$

$$= \frac{\frac{1}{4}\cos\theta|1\rangle + \frac{\sqrt{3}}{4}\cos\theta|2\rangle + \sin\theta|3\rangle}{\sqrt{\frac{1}{4} + \frac{3}{4}\sin^2\theta}},$$

because

$$\left|\left|\frac{1}{4}\cos\theta|1\rangle + \frac{\sqrt{3}}{4}\cos\theta|2\rangle + \sin\theta|3\rangle\right|\right|$$

$$= \sqrt{\left(\frac{1}{4}\cos\theta\right)^2 + \left(\frac{\sqrt{3}}{4}\cos\theta\right)^2 + (\sin\theta)^2}$$

$$= \sqrt{\frac{1}{4}\cos^2\theta + \sin^2\theta}$$

$$= \sqrt{\frac{1}{4} + \frac{3}{4}\sin^2\theta}.$$

(c) The expected value of $G$ in state $|\phi\rangle$ is

$$\langle\phi|G|\phi\rangle = \begin{pmatrix}\cos\theta & 0 & \sin\theta\end{pmatrix}\begin{pmatrix}\frac{5}{4} & \frac{\sqrt{3}}{4} & 0 \\ \frac{\sqrt{3}}{4} & \frac{7}{4} & 0 \\ 0 & 0 & 2\end{pmatrix}\begin{pmatrix}\cos\theta \\ 0 \\ \sin\theta\end{pmatrix}$$

$$= \begin{pmatrix}\cos\theta & 0 & \sin\theta\end{pmatrix}\begin{pmatrix}\frac{5}{4}\cos\theta \\ \frac{\sqrt{3}}{4}\cos\theta \\ 2\sin\theta\end{pmatrix}$$

$$= \frac{5}{4} \cos^2 \theta + 2 \sin^2 \theta$$

$$= \frac{5}{4} + \frac{3}{4} \sin^2 \theta.$$

The square of this is

$$\langle \phi | G | \phi \rangle^2 = \frac{25}{16} + \frac{15}{8} \sin^2 \theta + \frac{9}{16} \sin^4 \theta.$$

The operator $G^2$ is written as a matrix as

$$G^2 = \begin{pmatrix} \frac{5}{4} & \frac{\sqrt{3}}{4} & 0 \\ \frac{\sqrt{3}}{4} & \frac{7}{4} & 0 \\ 0 & 0 & 2 \end{pmatrix} \begin{pmatrix} \frac{5}{4} & \frac{\sqrt{3}}{4} & 0 \\ \frac{\sqrt{3}}{4} & \frac{7}{4} & 0 \\ 0 & 0 & 2 \end{pmatrix} = \begin{pmatrix} \frac{7}{4} & \frac{3\sqrt{3}}{4} & 0 \\ \frac{3\sqrt{3}}{4} & \frac{13}{4} & 0 \\ 0 & 0 & 4 \end{pmatrix},$$

so

$$\langle \phi | G^2 | \phi \rangle = \begin{pmatrix} \cos \theta & 0 & \sin \theta \end{pmatrix} \begin{pmatrix} \frac{7}{4} & \frac{3\sqrt{3}}{4} & 0 \\ \frac{2\sqrt{3}}{4} & \frac{13}{4} & 0 \\ 0 & 0 & 4 \end{pmatrix} \begin{pmatrix} \cos \theta \\ 0 \\ \sin \theta \end{pmatrix}$$

$$= \begin{pmatrix} \cos \theta & 0 & \sin \theta \end{pmatrix} \begin{pmatrix} \frac{7}{4} \cos \theta \\ \frac{3\sqrt{3}}{4} \cos \theta \\ 4 \sin \theta \end{pmatrix}$$

$$= \frac{7}{4} \cos^2 \theta + 4 \sin^2 \theta$$

$$= \frac{7}{4} + \frac{9}{4} \sin^2 \theta.$$

The dispersion is therefore

$$\Delta_\phi(G) = \sqrt{\frac{7}{4} + \frac{9}{4} \sin^2 \theta - \frac{25}{16} - \frac{15}{8} \sin^2 \theta - \frac{9}{16} \sin^4 \theta}$$

$$= \frac{1}{4} \sqrt{3 + 6 \sin^2 \theta - 9 \sin^4 \theta}$$

$$= \frac{1}{4} \sqrt{(3 \sin^2 \theta - 3)(3 \sin^2 \theta + 1)}.$$

(d) The dispersion is zero when $\sin \theta = \pm 1$, so when $\theta = \pm \pi/2, \pm 3\pi/2, \dots$ At these values of $\theta$ $\cos \theta = 0$ so the state $\phi$ is just a multiple of $|3\rangle$. We know from the earlier parts of the question that $|3\rangle$ is an eigenvector of $G$. As we know, the dispersion should indeed be zero in the state is an eigenvector of the operator in question.

8. (a) First, note that:

$$(H \mid \psi\rangle)^\dagger = (E \mid \psi\rangle)^\dagger,$$

from which it follows that:

$$\langle \psi \mid H^\dagger = \langle \psi \mid E^*.$$

Since $H$ is self-adjoint we have $H = H^\dagger$ and $E = E^*$, and therefore

$$\langle \psi \mid H = \langle \psi \mid E. \tag{3.12}$$

Now we turn to the main calculation:

$$\langle \psi \mid [H, A] \mid \psi\rangle = \langle \psi \mid HA - AH \mid \psi\rangle,$$
$$= \langle \psi \mid HA \mid \psi\rangle - \langle \psi \mid AH \mid \psi\rangle, \quad \text{using (3.12)},$$
$$= E\langle \psi \mid A \mid \psi\rangle - E\langle \psi \mid A \mid \psi\rangle = 0$$

(b)   (i) $[P^2, X] = P[P, X] + [P, X]P = P(-i\hbar) + (-i\hbar)P = -i\hbar 2P.$

     (ii)

$$[P^2, XP] = P[P, XP] + [P, XP]P,$$
$$= P\left(X[P, P] + [P, X]P\right)$$
$$+ \left(X[P, P] + [P, X]P\right)P,$$
$$= P\left(0 + (-i\hbar)P\right) + \left(0 + (-i\hbar)P\right)P,$$
$$= -2i\hbar P^2.$$

(iii)

$$[X^N, XP] = X[X^N, P] + [X^N, X]P = X[X^N, P].$$

We also have

$$[X^N, P] = [XX^{N-1}, P] = X[X^{N-1}, P] + [X, P]X^{N-1}$$
$$= X[X^{N-1}, P] + i\hbar X^{N-1}.$$

Iterating this same calculation gives:

$$[X^N, P] = X^{N-1}[X^{N-(N-1)}, P] + (N - 1)i\hbar X^{N-1}$$
$$= N i\hbar X^{N-1},$$

and therefore:

$$[X^N, XP] = X[X^N, P] = Ni\hbar X^N.$$

(c)

$$[H, X] = \frac{1}{2m}[P^2, X] + k[X^N, X]$$

$$\frac{1}{2m}(-i\hbar)2P + 0 = -i\hbar\frac{P}{m}, \quad \text{using part (b) (i).}$$

Therefore

$$\langle \psi \mid [H, X] \mid \psi \rangle = -\frac{i\hbar}{m}\langle \psi \mid P \mid \psi \rangle = 0, \quad \text{using part (a),}$$

or

$$\langle \psi \mid P \mid \psi \rangle = \langle P \rangle = 0.$$

(d)

$$[H, XP] = \frac{1}{2m}[P^2, XP] + k[X^N, XP],$$

$$= \frac{1}{2m}\left(X[P^2, P] + [P^2, X]P\right) + k[X^N, XP],$$

$$= \frac{1}{2m}\left(0 - i\hbar 2P^2\right) + kNi\hbar X^2.$$

using part (b) (i) and (iii).

Therefore we have:

$$\langle \psi \mid [H, XP] \mid \psi \rangle = -2i\hbar \left\langle \psi \mid \frac{P^2}{2m} \mid \psi \right\rangle + Ni\hbar \langle \psi \mid kX^N \mid \psi \rangle,$$

from which it follows that:

$$2\langle T \rangle = N\langle V \rangle.$$

(e) In part (c) we showed that $\langle P \rangle = 0$. In part (d) we showed that $\langle P^2 \rangle = 2m\langle T \rangle$. Therefore we have:

$$\langle (\Delta P^2) \rangle = \langle P^2 \rangle - \langle P \rangle^2 = 2m\langle T \rangle.$$

9. (a)

$$A = \begin{pmatrix} 2 & i \\ -i & 2 \end{pmatrix}.$$

(b) It is trivial to verify that $A = A^\dagger$.

(c)

$$\det\left(A - \lambda\mathbb{I}\right) = (2 - \lambda)^2 - 1 = \lambda^2 - 4\lambda + 3 = 0, \Rightarrow \lambda = 1,\, 3.$$

For $\lambda = 1$ the corresponding normalized eigenvector is computed as follows:

$$\begin{pmatrix} 2 & i \\ -i & 2 \end{pmatrix}\begin{pmatrix} a \\ b \end{pmatrix} = \begin{pmatrix} a \\ b \end{pmatrix}$$

which implies:

$$a + ib = 0$$

giving rise to the eigenvector:

$$\begin{pmatrix} a \\ b \end{pmatrix} = \begin{pmatrix} 1 \\ i \end{pmatrix},$$

which after normalization gives:

$$|e_1\rangle = \frac{1}{\sqrt{2}}\begin{pmatrix} 1 \\ i \end{pmatrix}.$$

For $\lambda = 3$ the corresponding normalized eigenvector is computed as follows:

$$\begin{pmatrix} 2 & i \\ -i & 2 \end{pmatrix}\begin{pmatrix} a \\ b \end{pmatrix} = 3\begin{pmatrix} a \\ b \end{pmatrix}$$

which implies:

$$-a + ib = 0,$$

giving rise to the eigenvector:

$$\begin{pmatrix} a \\ b \end{pmatrix} = \begin{pmatrix} -1 \\ i \end{pmatrix},$$

which after normalization gives:

$$|e_2\rangle = \frac{1}{\sqrt{2}}\begin{pmatrix} -1 \\ i \end{pmatrix}.$$

(d) It is straightforward to verify:

$$|1\rangle = \frac{1}{\sqrt{2}}|e_1\rangle - \frac{1}{\sqrt{2}}|e_2\rangle,$$

$$|2\rangle = \frac{1}{i\sqrt{2}}|e_1\rangle + \frac{1}{i\sqrt{2}}|e_2\rangle. \tag{3.13}$$

(e)

$$\langle e_1 \mid e_2 \rangle = \frac{1}{2} \left( 1 \;\; -i \right) \begin{pmatrix} -1 \\ i \end{pmatrix} = 0$$

(f)

$$A = 1\, |e_1\rangle\langle e_1| + 3\, |e_2\rangle\langle e_2|$$

(g)

$$A = 2\, |1\rangle\langle 1| + i\, |1\rangle\langle 2| - i\, |2\rangle\langle 1| 2\, |2\rangle\langle 2|$$

(h)

$$\mathrm{Pr}_{|1\rangle}(1) = \langle 1|e_1\rangle\langle e_1|1\rangle,$$
$$= \frac{1}{2}\left( \langle e_1| - \langle e_2| \right) |e_1\rangle\langle e_1| \left( |e_1\rangle - |e_2\rangle \right),$$
$$= \frac{1}{2}$$

(i)

$$\mathrm{Pr}_{|1\rangle}(3) = \langle 1|e_2\rangle\langle e_2|1\rangle,$$
$$= \frac{1}{2}\left( \langle e_1| - \langle e_2| \right) |e_2\rangle\langle e_2| \left( |e_1\rangle - |e_2\rangle \right),$$
$$= \frac{1}{2}$$

(j)

$$\mathrm{Pr}_{|2\rangle}(1) = \langle 2|e_1\rangle\langle e_1|2\rangle$$
$$= \left( \frac{1}{i\sqrt{2}} \right) \left( \frac{1}{-i\sqrt{2}} \right) \left( \langle e_1| + \langle e_2| \right)$$
$$\times |e_1\rangle\langle e_1| \left( |e_1\rangle + |e_2\rangle \right),$$
$$= \frac{1}{2}$$

(k)

$$\mathrm{Pr}_{|2\rangle}(3) = \langle 2|e_2\rangle\langle e_2|2\rangle$$
$$= \left( \frac{1}{i\sqrt{2}} \right) \left( \frac{1}{-i\sqrt{2}} \right)$$
$$\times \left( \langle e_1| + \langle e_2| \right) |e_2\rangle\langle e_2| \left( |e_1\rangle + |e_2\rangle \right),$$
$$= \frac{1}{2}$$

(l) If the result of the measurement of $A$ in the state $|1\rangle$ is 1, then the state of the system after measurement is given by:

$$\frac{|e_1\rangle\langle e_1|1\rangle}{\sqrt{\langle 1|e_1\rangle\langle e_1|1\rangle}}.$$

Using (3.13) we obtain:

$$\langle e_1|1\rangle = \langle e_1|\left(\frac{1}{\sqrt{2}}|e_1\rangle - \frac{1}{\sqrt{2}}|e_2\rangle\right) = \frac{1}{\sqrt{2}},$$

and therefore:

$$\frac{|e_1\rangle\langle e_1|1\rangle}{\sqrt{\langle 1|e_1\rangle\langle e_1|1\rangle}} = |e_1\rangle,$$

i.e. the eigenstate corresponding to the eigenvalue that was measured, as expected.

(m) If the result of the measurement of $A$ in the state $|1\rangle$ is 3, then the state of the system after measurement is given by:

$$\frac{|e_2\rangle\langle e_2|1\rangle}{\sqrt{\langle 1|e_2\rangle\langle e_2|1\rangle}}$$

Using (3.13) we obtain:

$$\langle e_2|1\rangle = \langle e_2|\left(\frac{1}{\sqrt{2}}|e_1\rangle - \frac{1}{\sqrt{2}}|e_2\rangle\right) = -\frac{1}{\sqrt{2}},$$

and therefore

$$\frac{|e_2\rangle\langle e_2|1\rangle}{\sqrt{\langle 1|e_2\rangle\langle e_2|1\rangle}} = -|e_2\rangle,$$

i.e. the eigenstate corresponding to the eigenvalue that was measured, multiplied by an irrelevant phase factor $-1$.

(n) If the result of the measurement of $A$ in the state $|2\rangle$ is 1, then the state of the system after measurement is given by:

$$\frac{|e_1\rangle\langle e_1|2\rangle}{\sqrt{\langle 2|e_1\rangle\langle e_1|2\rangle}},$$

Using (3.13) we obtain:

$$\langle e_1|2\rangle = \langle e_1|\left(\frac{1}{i\sqrt{2}}|e_1\rangle + \frac{1}{i\sqrt{2}}|e_2\rangle\right) = \frac{1}{\sqrt{2}} = \frac{1}{i\sqrt{2}},$$

and therefore

$$\frac{|e_1\rangle\langle e_1|2\rangle}{\sqrt{\langle 2|e_1\rangle\langle e_1|2\rangle}} = -i|e_1\rangle,$$

i.e. the eigenstate corresponding to the eigenvalue that was measured, multiplied by an irrelevant phase factor $-i$.

(o) If the result of the measurement of $A$ in the state $|2\rangle$ is 3, then the state of the system after measurement is given by:

$$\frac{|e_2\rangle\langle e_2|2\rangle}{\sqrt{\langle 2|e_2\rangle\langle e_2|2\rangle}}$$

Using (3.13) we obtain:

$$\langle e_2|2\rangle = \langle e_2| \left( \frac{1}{i\sqrt{2}}|e_1\rangle + \frac{1}{i\sqrt{2}}|e_2\rangle \right) = \frac{1}{\sqrt{2}} = \frac{1}{i\sqrt{2}},$$

and therefore

$$\frac{|e_2\rangle\langle e_2|2\rangle}{\sqrt{\langle 2|e_2\rangle\langle e_2|2\rangle}} = -i|e_2\rangle,$$

i.e. the eigenstate corresponding to the eigenvalue that was measured, multiplied by an irrelevant phase factor $-i$.

10. (a) $P$ is self-adjoint, $X$ is self-adjoint, since $m\omega$ is real, $m\omega X$ is self-adjoint. Therefore $a = P + m\omega X$ is self-adjoint. By the same reasoning $b = P - m\omega X$ is self-adjoint.

(b) $N$ is not self-adjoint since $a$ and $b$ do not commute.

$$N^\dagger = \frac{(ba)^\dagger}{2m\hbar\omega} = \frac{a^\dagger b^\dagger}{2m\hbar\omega} = \frac{ab}{2m\hbar\omega} \neq N.$$

(c) $[N, a] = \frac{1}{2m\hbar\omega}[ba, a] = \frac{1}{2m\hbar\omega}(b[a, a] + [b, a]a)$. We have

$$[b, a] = [P - m\omega X, P + m\omega X] = -2im\hbar\omega.$$

Therefore

$$[N, a] = -ia.$$

(d) $[N, b] = \frac{1}{2m\hbar\omega}[ba, b] = \frac{1}{2m\hbar\omega}(b[a, b] + [b, b]a)$. Using the result from part (c) for $-[b, a] = [a, b]$ we have

$$[N, b] = ib.$$

(e) $ba = (P - m\omega X)(P + m\omega X) = P^2 - m\omega(XP - PX) - m^2\omega^2 X^2$.
Therefore,

$$ba + mi\hbar\omega = P^2 - m^2\omega^2 X^2 = 2mH.$$

From this expression we obtain,

$$\frac{ba}{2m} + \frac{i}{2}\hbar\omega = H,$$

or,

$$H = \hbar\omega\left(\frac{ba}{2m\hbar\omega} + \frac{i}{2}\right) = \hbar\omega\left(N + \frac{i}{2}\right).$$

11. (a)

$$\langle x \rangle = \int_{-\infty}^{\infty} \bar{\psi}x\psi\,dx = 0,$$

since $x$ is odd and $\psi^2$ is even.

(b)

$$\langle x^2 \rangle = \int_{-\infty}^{\infty} \bar{\psi}x^2\psi\,dx,$$

$$= \left(\frac{2a}{\pi}\right)^{\frac{1}{2}} \int_{-\infty}^{\infty} x^2 e^{-2ax^2}\,dx = \frac{1}{4a}.$$

(c)

$$\langle p \rangle = \int_{-\infty}^{\infty} \bar{\psi}p\psi\,dx,$$

$$= \left(\frac{2a}{\pi}\right)^{\frac{1}{2}} \int_{-\infty}^{\infty} e^{-ax^2}\left(\frac{\hbar}{i}\frac{d}{dx}\right)e^{-ax^2}\,dx,$$

$$= \left(\frac{2a}{\pi}\right)^{\frac{1}{2}}\frac{\hbar}{i}(-2a)\int_{-\infty}^{\infty} xe^{-2ax^2}\,dx = 0,$$

for the same reason as part (a).

(d)

$$\langle p^2 \rangle = \int_{-\infty}^{\infty} \bar{\psi} p^2 \psi \, dx,$$

$$= -\hbar^2 \left( \frac{2a}{\pi} \right)^{\frac{1}{2}} \int_{-\infty}^{\infty} e^{-ax^2} \left( \frac{d^2}{dx^2} \right) e^{-ax^2} \, dx,$$

$$= -\hbar^2 \left( \frac{2a}{\pi} \right)^{\frac{1}{2}} \int_{-\infty}^{\infty} 2a(2ax^2 - 1)e^{-2ax^2} \, dx,$$

$$= -4\hbar^2 a^2 \left( \frac{2a}{\pi} \right)^{\frac{1}{2}} \int_{-\infty}^{\infty} x^2 e^{-2ax^2} \, dx + 2\hbar^2 a \left( \frac{2a}{\pi} \right)^{\frac{1}{2}}$$

$$\times \int_{-\infty}^{\infty} e^{-2ax^2} \, dx = \hbar^2 a.$$

(e) Assembling the pieces we have already computed:

$$\Delta x^2 = \langle x^2 \rangle - \langle x \rangle^2 = \frac{1}{4a},$$

$$\Delta x = \frac{1}{2\sqrt{a}},$$

$$\Delta p^2 = \langle p^2 \rangle - \langle p \rangle^2 = a\hbar^2,$$

$$\Delta p = \sqrt{a}\hbar,$$

we obtain

$$\Delta p \Delta x = \sqrt{a}\hbar \frac{1}{2\sqrt{a}} = \frac{\hbar}{2}.$$

12. In this problem we repeatedly use the following commutator identity from class:

$$[AB, C] = A[B, C] + [A, C]B$$

(a) Ehrenfest's equations have the form:

$$\frac{d}{dt}\langle P \rangle = 0,$$

$$\frac{d}{dt}\langle X \rangle = \frac{1}{i\hbar}\langle [X, cP] \rangle = c,$$

with solution:

$$\langle X \rangle_t = x_0 + ct,$$

$$\langle P \rangle_t = p_0 = \text{constant}.$$

(b) We have:

$$\sigma_p^2(t) = \langle (P - \langle P\rangle_t)^2 \rangle,$$
$$= \langle P^2 \rangle_t - (\langle P\rangle_t)^2.$$

We have already shown that $\langle P\rangle_t$ is constant. We also have:

$$\frac{d}{dt}\langle P^2\rangle_t = \frac{c}{2i\hbar m}\langle [P^2, P]\rangle_t = 0.$$

Hence, $\langle P^2\rangle_t$ is constant in time, and the result follows.

(c) We have:

$$\frac{d}{dt}\langle X^2\rangle_t = \frac{1}{i\hbar}\langle [X^2, cP]\rangle_t = 2c\langle X\rangle_t$$
$$= 2c(x_0 + ct).$$

(d) Integrating the expression from part (c) gives:

$$\langle X^2\rangle_t = s + 2cx_0 t + c^2 t^2,$$

where $s$ is a constant.

(e) Using the expression from part (d) and the solution of Ehrenfest's equations gives:

$$\langle X^2\rangle_t - \langle X\rangle_t^2 = s + 2cx_0 t + c^2 t^2 - (x_0^2 + 2x_0 ct + c^2 t^2),$$
$$= s - x_0^2 = \text{constant}.$$

13. (a) We apply the product rule for differentiation:

$$\frac{d}{dt}\langle\psi|A|\psi\rangle = \left(\frac{\partial}{\partial t}\langle\psi|\right)A|\psi\rangle + \langle\psi|\frac{\partial A}{\partial t}|\psi\rangle + \langle\psi|A\left(\frac{\partial|\psi\rangle}{\partial t}\right).$$

For the terms involving partial derivatives with respect to $t$ we insert the expressions from the Schrödinger equattion and its complex conjugate:

$$\frac{\partial}{\partial t}|\psi\rangle = \frac{1}{i\hbar}H|\psi\rangle,$$
$$\frac{\partial}{\partial t}\langle\psi| = -\frac{1}{i\hbar}\langle\psi|H,$$

therefore obtaining:

$$= -\frac{1}{i\hbar}\langle\psi|HA|\psi\rangle + \langle\psi|\frac{\partial A}{\partial t}|\psi\rangle + \langle\psi|A\frac{1}{i\hbar}H|\psi\rangle,$$

$$= \frac{1}{i\hbar}\langle\psi|AH - HA|\psi\rangle + \langle\psi|\frac{\partial A}{\partial t}|\psi\rangle,$$

$$= \frac{1}{i\hbar}\langle\psi|[A, H]|\psi\rangle + \langle\psi|\frac{\partial A}{\partial t}|\psi\rangle, \tag{3.14}$$

which is the desired result.

(b) We apply (3.14) to the operators $X$ and $P$. For $X$ we have:

$$\frac{d}{dt}\langle X\rangle = \frac{1}{i\hbar}\langle[X, H]\rangle.$$

It is clear that $[X, V(X)] = 0$, from which it follows that $[X, H]$ reduces to:

$$\left[X, \frac{P^2}{2m}\right] = \frac{P}{2m}[X, P] + [X, P]\frac{P}{2m},$$

$$= \frac{P}{2m}i\hbar + i\hbar\frac{P}{2m} = i\hbar\frac{P}{m}.$$

Therefore we have:

$$\frac{d}{dt}\langle X\rangle = \frac{\langle P\rangle}{m}.$$

Now we apply (3.14) to $P$:

$$\frac{d}{dt}\langle P\rangle = \frac{1}{i\hbar}\langle[P, H]\rangle,$$

$$= \frac{1}{i\hbar}\langle[P, V(X)]\rangle,$$

$$= \frac{1}{i\hbar}(-i\hbar)\left\langle\frac{dV}{dx}(X)\right\rangle.$$

Using the result:

$$[P, X^n] = -i\hbar n X^{n-1},$$

we obtain:

$$\frac{d}{dt}\langle P\rangle = -\left\langle\frac{dV}{dx}(X)\right\rangle.$$

(c) This would be true if:

$$\left\langle \frac{dV}{dx}(X) \right\rangle = \frac{d}{dx}V(\langle X \rangle),$$

but this equality does not generally hold.

14. (a) $|\langle \psi_1 \mid \phi_n \rangle|^2$ is the probability of measuring $\alpha_n$ in the state $|\psi_1\rangle$ and $|\langle \psi_2 \mid \phi_n \rangle|^2$ is the probability of measuring $\alpha_n$ in the state $|\psi_2\rangle$

(b)

$$\langle \psi | \psi \rangle = (3\langle \psi_1 | + 4i\langle \psi_2 |)(|\psi\rangle = 3|\psi_1\rangle - 4i|\psi_2\rangle),$$

$$= 9\langle \psi_1 | \psi_1 \rangle + 16\langle \psi_2 | \psi_2 \rangle = 25 = \|\psi\|^2$$

Hence the normalized $|\psi\rangle$

$$|\psi\rangle = \frac{3|\psi_1\rangle - 4i|\psi_2\rangle}{5}$$

(c)

$$\text{Prob}_{\alpha_n}(|\psi\rangle) = \langle \psi | \phi_n \rangle \overline{\langle \psi | \phi_n \rangle},$$

$$= \langle \psi | \phi_n \rangle \langle \phi_n | \psi \rangle,$$

$$= \frac{1}{25}\left(((3\langle \psi_1 | + 4i\langle \psi_2 |)|\phi_n\rangle)\right)$$

$$\times \left((\langle \phi_n |(3|\psi_1\rangle - 4i|\psi_2\rangle)))\right),$$

$$= \frac{1}{25}\left(9\langle \psi_1 | \phi_n \rangle\langle \phi_n | \psi_1 \rangle - 12i\langle \psi_1 | \phi_n \rangle\langle \phi_n | \psi_2 \rangle\right.$$

$$\left. + 12i\langle \psi_2 | \phi_n \rangle\langle \phi_n | \psi_1 \rangle + 16\langle \psi_2 | \phi_n \rangle\langle \phi_n | \psi_2 \rangle\right),$$

$$= 9\langle \psi_1 | \phi_n \rangle\langle \phi_n | \psi_1 \rangle + 16\langle \psi_2 | \phi_n \rangle\langle \phi_n | \psi_2 \rangle$$

$$+ 2\text{Re}\left(12i\langle \psi_1 | \phi_n \rangle\langle \phi_n | \psi_2 \rangle\right)$$

(d) The state of the system after $\alpha_n$ is measured is:

$$|\phi_n\rangle\langle \phi_n | \left(\frac{3|\psi_1\rangle - 4i|\psi_2\rangle}{5}\right) = \left(\frac{3}{5}\langle \phi_n | \psi_1 \rangle - \frac{4i}{5}\langle \phi_n | \psi_2 \rangle\right)|\phi_n\rangle,$$

The normalized state is $|\phi_n\rangle$.

15. (a)

$$E_{\psi_n}(X) = \frac{2}{a} \int_0^a x \sin^2\left(\frac{n\pi x}{a}\right) dx = \frac{a}{2}$$

(b)

$$E_{\psi_n}(X^2) = \frac{2}{a} \int_0^a x^2 \sin^2\left(\frac{n\pi x}{a}\right) dx = a^2\left(\frac{1}{3} - \frac{1}{2n^2\pi^2}\right).$$

(c)

$$E_{\psi_n}(P) = \frac{2}{a} \int_0^a \left(\sin\left(\frac{n\pi x}{a}\right)\right) \frac{\hbar}{i}\frac{d}{dx}\left(\sin\left(\frac{n\pi x}{a}\right)\right) dx,$$

$$= \frac{2n\pi\hbar}{ia^2} \int_0^a \sin\left(\frac{n\pi x}{a}\right) \cos\left(\frac{n\pi x}{a}\right) dx = 0.$$

(d)

$$E_{\psi_n}(P^2) = \frac{2}{a} \int_0^a \left(\sin\left(\frac{n\pi x}{a}\right)\right)\left(-\hbar^2\frac{d^2}{dx^2}\right)\left(\sin\left(\frac{n\pi x}{a}\right)\right) dx,$$

$$= \frac{2\hbar^2 n^2\pi^2}{a^3} \int_0^a \sin^2\left(\frac{n\pi x}{a}\right) dx = \frac{\hbar^2 n^2\pi^2}{a^2}$$

(e) Using the results from above, we have,

$$\Delta_{\psi_n}(X)^2 = E_{\psi_n}(X^2) - (E_{\psi_n}(X))^2 = \frac{n^2\pi^2 - 6}{12n^2\pi^2}a^2,$$

and

$$\Delta_{\psi_n}(P)^2 = E_{\psi_n}(P^2) - (E_{\psi_n}(P))^2 = \frac{\hbar^2 n^2\pi^2}{a^2}.$$

and therefore the uncertainty relation is given as follows:

$$\Delta_{\psi_n}(X)\Delta_{\psi_n}(P) = \hbar\sqrt{\frac{n^2\pi^2 - 6}{12}}.$$

You can see by inspection that this quantity is smallest when $n = 1$.

16. For the calculations we will need several identities. First, we write $X$ and $P$ in terms of $a$ and $a^\dagger$:

$$X = \frac{1}{2im\omega}(a^\dagger - a),$$

and

$$P = \frac{1}{2}(a^\dagger + a).$$

From these expressions we obtain:

$$X^2 = -\frac{1}{4m^2\omega^2}(a^\dagger - a)(a^\dagger - a),$$

$$= -\frac{1}{4m^2\omega^2}(a^\dagger a^\dagger - a^\dagger a - aa^\dagger + aa),$$

and

$$P^2 = \frac{1}{4}(a^\dagger + a)(a^\dagger + a),$$

$$= \frac{1}{4}(a^\dagger a^\dagger + a^\dagger a + aa^\dagger + aa).$$

From the identities given in the statement of the question, we also have:

$$\langle n|aa^\dagger|n\rangle = \langle a^\dagger n|a^\dagger|n\rangle = 2m\hbar\omega(n+1), \qquad (\star)$$

and

$$\langle n|a^\dagger a|n\rangle = \langle an|a|n\rangle = 2m\hbar\omega n. \qquad (\star\star)$$

(a) Using orthogonality of $|n\rangle$ and the expression for $X$ in terms of $a$ and $a^\dagger$ we conclude that $\langle n|X|n\rangle = 0$.

(b) As in part (a), but using the expression for $P$ is terms of $a$ and $a^\dagger$ we obtain $\langle n|P|n\rangle = 0$.

(c) We use the expression for $X^2$ in terms of $a$ and $a^\dagger$, orthogonality of $|n\rangle$, and the identities $(\star)$ and $(\star\star)$ to obtain:

$$\langle n|X^2|n\rangle = \frac{1}{4m^2\omega^2}\langle n|a^\dagger a + aa^\dagger|n\rangle,$$

$$= \frac{1}{4m^2\omega^2}(2m\hbar\omega n + 2m\hbar\omega(n+1)),$$

$$= \frac{\hbar}{2m\omega}(2n+1).$$

(d) Similar to part (c), except we use the expression for $P^2$ in terms of $a$ and $a^\dagger$, orthogonality of $|n\rangle$, and the identities $(\star)$ and $(\star\star)$ to obtain:

$$\langle n|P^2|n\rangle = \frac{1}{4}\langle n|a^\dagger a + aa^\dagger|n\rangle,$$

$$= \frac{m\hbar\omega}{2}(2n+1).$$

(e) Recall

$$\Delta X = \sqrt{\langle X^2\rangle - \langle X\rangle^2},$$

and

$$\Delta P = \sqrt{\langle P^2\rangle - \langle P\rangle^2},$$

and we have shown $\langle X\rangle = \langle P\rangle = 0$. Therefore we have:

$$\Delta P \Delta X = \sqrt{\langle P^2\rangle}\sqrt{\langle X^2\rangle},$$

$$= \sqrt{\left(\frac{\hbar}{2m\omega}\frac{m\hbar\omega}{2}\right)(2n+1)(2n+1)},$$

$$= \frac{\hbar}{2}(2n+1),$$

and this quantity is the smallest for $n = 0$, the ground state.

17. (a) We use the expression for the Hamiltonian in terms of the number operator, $H = \hbar\omega(N + \frac{1}{2})$ to obtain:

$$\hat{a}(t) \equiv e^{\frac{i}{\hbar}Ht} a\, e^{-\frac{i}{\hbar}Ht},$$

$$= e^{\frac{i}{\hbar}t(\hbar\omega(N+\frac{1}{2}))} a e^{-\frac{i}{\hbar}t(\hbar\omega(N+\frac{1}{2}))},$$

$$= e^{i\omega t N} e^{\frac{i\omega t}{2}} a e^{\frac{-i\omega t}{2}} e^{-i\omega t N},$$

$$= e^{i\omega t N} a e^{-i\omega t N}$$

Note that $e^{\frac{-i\omega t}{2}}$ and $e^{\frac{i\omega t}{2}}$ are scalars. Hence they can be "moved around" in the expression (by linearity). Therefore they combine to give the number one, and we obtain the final line, which was the result that we sought.

(b) Noting that $[N, a] = -a$, and using the result from part (a), we have:

$$\frac{d}{dt}\hat{a}(t) = \frac{d}{dt}\left(e^{i\omega t N} a e^{-i\omega t N}\right),$$

$$= \frac{d}{dt}\left(e^{i\omega t N}\right) a e^{-i\omega t N} + e^{i\omega t N} a \frac{d}{dt}\left(e^{-i\omega t N}\right),$$

$$= e^{i\omega N t}(i\omega N) a e^{-i\omega t N} + e^{i\omega N t} a(-i\omega N) e^{-i\omega N t},$$

$$= i\omega e^{i\omega N t}(Na - aN) e^{-i\omega N t},$$

$$= -i\omega e^{i\omega N t} a e^{-i\omega N t},$$

$$= -i\omega \hat{a}.$$

(c) The initial value is obtained from the expression in part (a)

$$\frac{d}{dt}\hat{a}(t) = -i\omega\hat{a}, \quad \hat{a}(0) = a.$$

The solution of this ordinary differential equation is:

$$\hat{a}(t) = e^{-i\omega t} a.$$

(d) We take the adjoint of the expression in part (c) to obtain:

$$\hat{a}^\dagger(t) = e^{i\omega t} a^\dagger$$

(e) Using $X = \frac{1}{2im\omega}(a^\dagger - a)$ and using

$$\hat{a}(t) \equiv e^{\frac{i}{\hbar}Ht} a\, e^{-\frac{i}{\hbar}Ht},$$

$$\hat{X}(t) \equiv e^{\frac{i}{\hbar}Ht} X\, e^{-\frac{i}{\hbar}Ht},$$

We have:

$$\hat{X}(t) \equiv \frac{1}{2im\omega} e^{\frac{i}{\hbar}Ht}\left(a^\dagger - a\right) e^{-\frac{i}{\hbar}Ht},$$

$$= \frac{1}{2im\omega}(\hat{a}^\dagger - \hat{a}).$$

18. Recall that for a general operator $A$ we have

$$\frac{d}{dt}\langle A\rangle = \frac{1}{i\hbar}\langle[A, H]\rangle + \left\langle\frac{\partial A}{\partial t}\right\rangle, \qquad (\dagger).$$

This formula is used in the solution to this problem. In this problem the operator is independent of time and the last expression vanishes. For operators $A$, $B$, $C$ we also repeatedly use the commutator relation:

$$[AB, C] = A[B, C] + [A, C]B, \qquad (\ddagger).$$

(a) Ehrenfest's equations are given by:

$$\frac{d}{dt}\langle X \rangle = \frac{\langle P \rangle}{m},$$

$$\frac{d}{dt}\langle P \rangle = 0,$$

with solutions

$$\langle X \rangle_t = \frac{\langle P \rangle_0}{m}t + \langle X \rangle_0,$$

$$\langle P \rangle_t = \langle P \rangle_0,$$

where $\langle \cdot \rangle_t$ denotes the expectation value at time $t$.

(b) Using (†) we have:

$$\frac{d}{dt}\langle P^2 \rangle = \frac{1}{2mi\hbar}\langle [P^2, P^2] \rangle = 0.$$

Hence $\langle P^2 \rangle$ is constant in time.

(c) Using (†) we have:

$$\frac{d}{dt}\langle (XP + PX) \rangle = \frac{1}{2mi\hbar}\langle [XP + PX, P^2] \rangle.$$

We can simplify the right hand side using ($\ddagger$) and the commutation relation $[X, P] = i\hbar$ to obtain:

$$\begin{aligned}
[XP + PX, P^2] &= P[XP + PX, P] + [XP + PX]P, \\
&= P\left([XP, P] + [PX, P]\right) \\
&\quad + \left([XP, P] + [PX, P]\right)P, \\
&= P^2\,(4i\hbar),
\end{aligned}$$

and therefore:

$$\frac{d}{dt}\langle (XP + PX) \rangle = \frac{2}{m}\langle P^2 \rangle.$$

(d) Using (†) we have:

$$\frac{d}{dt}\langle X^2 \rangle = \frac{1}{2mi\hbar}\langle [X^2, P^2] \rangle.$$

We can simplify the right hand side using (‡) and the commutation relation $[X, P] = i\hbar$ to obtain:

$$[X^2, P^2] = X[X, P^2] + [X, P^2]X,$$
$$= X\left(P[X, P] + [X, P]P\right) + \left(P[X, P] + [X, P]P\right)X,$$
$$= 2i\hbar(XP + PX),$$

and therefore:

$$\frac{d}{dt}\langle X^2 \rangle = \frac{1}{m}(XP + PX).$$

(e) We have shown that $\langle P^2 \rangle$ is constant and $\langle P \rangle$ is constant. Therefore:

$$\langle P^2 \rangle - \langle P \rangle^2 = \text{constant}.$$

## Chapter 4

1. In Chapter 3, Problem 1, we proved:

$$[AB, C] = ABC - CAB$$
$$= ABC - ACB + ACB - CAB$$
$$= A[B, C] + [A, C]B.$$

Similarly,

$$[AB, CD] = A[B, CD] + [A, CD]B$$
$$= AC[B, D] + A[B, C]D + C[A, D]B + [A, C]DB.$$

(Note, this is also equal to

$$CA[B, D] + C[A, D]B + A[B, C]D + [A, C]BD,$$

if you expand the second argument in the commutator first.)

So,

$$[L_j, L_k] = \left[\sum_{s,r} \epsilon_{jsr} X_s P_r, \sum_{t,u} \epsilon_{ktu} X_t P_u\right]$$

$$= \sum_{s,r,t,u} \epsilon_{jsr}\epsilon_{ktu}[X_s P_r, X_t P_u]$$

$$= \sum_{s,r,t,u} \epsilon_{jsr}\epsilon_{ktu}(X_s X_t[P_r, P_u] + X_s[P_r, X_t]P_u$$

$$+ X_t[X_s, P_u]P_r + [X_s, X_t]P_u P_r)$$

$$= \sum_{s,r,t,u} \epsilon_{jsr}\epsilon_{ktu}(-i\hbar X_s P_u \delta_{rt} + i\hbar X_t P_r \delta_{su})$$

$$= \sum_{s,t,u} \epsilon_{jst}\epsilon_{ktu}(-i\hbar X_s P_u) + \sum_{s,r,t} \epsilon_{jsr}\epsilon_{kts}(i\hbar X_t P_r)$$

$$= \sum_{s,u} \left(\sum_t \epsilon_{jst}\epsilon_{ukt}\right)(-i\hbar X_s P_u) + \sum_{r,t} \left(\sum_s \epsilon_{rjs}\epsilon_{kts}\right)(i\hbar X_t P_r).$$

For fixed $r, k, j, t$ we have the relation

$$\sum_s \epsilon_{rjs}\epsilon_{kts} = \delta_{rk}\delta_{jt} - \delta_{rt}\delta_{jk}.$$

Thus

$$[L_j, L_k] = \sum_{s,u} (\delta_{ju}\delta_{sk} - \delta_{jk}\delta_{su})(-i\hbar X_s P_u)$$

$$+ \sum_{r,t} (\delta_{rk}\delta_{jt} - \delta_{rt}\delta_{jk})(i\hbar X_t P_r)$$

$$= -i\hbar X_k P_j + i\hbar\delta_{jk}\sum_s X_s P_s + i\hbar X_j P_k - i\hbar\delta_{jk}\sum_t X_t P_t$$

$$= -i\hbar X_k P_j + i\hbar X_j P_k$$

$$= i\hbar \sum_m \epsilon_{jkm} \sum_{u,v} \epsilon_{muv} X_u P_v$$

$$= i\hbar \sum_m \epsilon_{jkm} L_m.$$

2. (a)
$$[J^2, J_\pm] = [J^2, J_1 \pm iJ_2]$$
$$= [J^2, J_1] \pm i[J^2, J_2] = 0$$

(b)
$$J_+ J_- = (J_1 + iJ_2)(J_1 - iJ_2)$$
$$= J_1^2 + iJ_2 J_1 - iJ_1 J_2 + J_2^2$$
$$= J^2 - J_3^2 - i[J_1, J_2]$$
$$= J^2 - J_3^2 + \hbar J_3$$

(c)
$$J_- J_+ = (J_1 - iJ_2)(J_1 + iJ_2)$$
$$= J_1^2 - iJ_2 J_1 + iJ_1 J_2 + J_2^2$$
$$= J^2 - J_3^2 + i[J_1, J_2]$$
$$= J^2 - J_3^2 - \hbar J_3$$

(d)
$$[J_+, J_-] = J_+ J_- - J_- J_+$$
$$= J^2 - J_3^2 + \hbar J_3 - J^2 + J_3^2 + \hbar J_3$$
$$= 2\hbar J_3$$

(e)
$$[J_3, J_\pm] = [J_3, J_1 \pm iJ_2]$$
$$= [J_3, J_1] \pm i[J_3, J_2]$$
$$= i\hbar J_2 \pm i(-i\hbar J_1)$$
$$= i\hbar J_2 \pm \hbar J_1$$
$$= \pm\hbar(J_1 \pm iJ_2) = \pm\hbar J_\pm$$

3. (a)
$$\sigma_x^2 = \begin{pmatrix} 0 & 1 \\ 1 & 0 \end{pmatrix} \begin{pmatrix} 0 & 1 \\ 1 & 0 \end{pmatrix} = \begin{pmatrix} 1 & 0 \\ 0 & 1 \end{pmatrix}$$

$$\sigma_y^2 = \begin{pmatrix} 0 & -i \\ i & 0 \end{pmatrix} \begin{pmatrix} 0 & -i \\ i & 0 \end{pmatrix} = \begin{pmatrix} 1 & 0 \\ 0 & 1 \end{pmatrix}$$

$$\sigma_z \sigma_x = \begin{pmatrix} 1 & 0 \\ 0 & -1 \end{pmatrix} \begin{pmatrix} 0 & 1 \\ 1 & 0 \end{pmatrix} = \begin{pmatrix} 0 & 1 \\ -1 & 0 \end{pmatrix} = i \begin{pmatrix} 0 & -i \\ i & 0 \end{pmatrix} = i\sigma_y$$

$$\sigma_z \sigma_y = \begin{pmatrix} 1 & 0 \\ 0 & -1 \end{pmatrix} \begin{pmatrix} 0 & -i \\ i & 0 \end{pmatrix} = \begin{pmatrix} 0 & -i \\ -i & 0 \end{pmatrix} = -i\sigma_x$$

(b)

$$A \cdot \sigma = A_x \sigma_x + A_y \sigma_y + A_z \sigma_z$$

$$= \begin{pmatrix} 0 & A_x \\ A_x & 0 \end{pmatrix} + \begin{pmatrix} 0 & -iA_y \\ iA_y & 0 \end{pmatrix} + \begin{pmatrix} A_z & 0 \\ 0 & -A_z \end{pmatrix}$$

$$= \begin{pmatrix} A_z & A_x - iA_y \\ A_x + iA_y & -A_z \end{pmatrix}.$$

Similarly

$$B \cdot \sigma = \begin{pmatrix} B_z & B_x - iB_y \\ B_x + iB_y & -B_z \end{pmatrix}.$$

So,

$$(A \cdot \sigma)(B \cdot \sigma)$$

$$= \begin{pmatrix} A_z & A_x - iA_y \\ A_x + iA_y & -A_z \end{pmatrix} \begin{pmatrix} B_z & B_x - iB_y \\ B_x + iB_y & -B_z \end{pmatrix}$$

$$= \begin{pmatrix} A_z B_z + A_x B_x + A_y B_y - iA_y B_x + iA_x B_y \\ A_x B_z + iA_y B_z - A_z B_x - iA_z B_y \end{pmatrix}$$

$$\begin{pmatrix} A_z B_x - iA_z B_y - A_x B_z + iA_y B_z \\ A_x B_x + A_y B_y + A_z B_z + iA_y B_x - iA_x B_y \end{pmatrix}$$

$$= (A_x B_x + A_y B_y + A_z B_z)I + i(A_y B_z - A_z B_y)\sigma_x$$

$$+ i(-A_x B_z + A_z B_x)\sigma_y + i(A_x B_y - A_y B_x)\sigma_z$$

$$= (A \cdot B)I + i(A \times B)\sigma.$$

†

4. In the $j = 1$ representation a basis of orthonormal vectors is $|1 \ -1\rangle$, $|1 \ 0\rangle$ and $|1 \ 1\rangle$.

$$J_3|1 \ 1\rangle = \hbar|1 \ 1\rangle$$

$$J_3|1 \ 0\rangle = 0$$

$$J_3|1 \ -1\rangle = -\hbar|1 \ -1\rangle$$

$$J_+|1 \ 1\rangle = \hbar\sqrt{1(1+1) - 1(1+1)} = 0$$

$$J_+|1 \ 0\rangle = \hbar\sqrt{1(1+1) - 0(0+1)}|1 \ 1\rangle = \hbar\sqrt{2}|1 \ 1\rangle$$

$$J_+|1 \ -1\rangle = \hbar\sqrt{1(1+1) + 1(-1+1)}|1 \ 0\rangle = \hbar\sqrt{2}|1 \ 0\rangle$$

$$J_-|1\ 1\rangle = \hbar\sqrt{1(1+1)-1(1-1)} = \hbar\sqrt{2}|1\ 0\rangle$$

$$J_-|1\ 0\rangle = \hbar\sqrt{1(1+1)-0(0-1)}|1\ 1\rangle = \hbar\sqrt{2}|1\ -1\rangle$$

$$J_-|1\ -1\rangle = 0$$

So,

$$J_1|1\ 1\rangle = \tfrac{1}{2}(J_+ + J_-)|1\ 1\rangle = \frac{\hbar}{\sqrt{2}}|1\ 0\rangle$$

$$J_1|1\ 0\rangle = \tfrac{1}{2}(J_+ + J_-)|1\ 0\rangle = \frac{\hbar}{\sqrt{2}}|1\ 1\rangle + \frac{\hbar}{\sqrt{2}}|1\ -1\rangle$$

$$J_1|1\ -1\rangle = \tfrac{1}{2}(J_+ + J_-)|1\ -1\rangle = \frac{\hbar}{\sqrt{2}}|1\ 0\rangle$$

$$J_2|1\ 1\rangle = \frac{-i}{2}(J_+ - J_-)|1\ 1\rangle = \frac{i\hbar}{\sqrt{2}}|1\ 0\rangle$$

$$J_2|1\ 0\rangle = \frac{-i}{2}(J_+ - J_-)|1\ 0\rangle = \frac{-i\hbar}{\sqrt{2}}|1\ 1\rangle + \frac{i\hbar}{\sqrt{2}}|1-1\rangle$$

$$J_2|1\ -1\rangle = \frac{-i}{2}(J_+ - J_-)|1\ -1\rangle = \frac{-i\hbar}{\sqrt{2}}|1\ 0\rangle$$

Now we can check the commutation relations:

$$[J_1, J_2]|1\ 1\rangle = J_1 J_2|1\ 1\rangle - J_2 J_1|1\ 1\rangle$$

$$= \frac{i\hbar}{\sqrt{2}} J_1|1\ 0\rangle - \frac{\hbar}{\sqrt{2}} J_2|1\ 0\rangle$$

$$= \frac{i\hbar^2}{2}|1\ 1\rangle + \frac{i\hbar^2}{2}|1\ -1\rangle + \frac{i\hbar^2}{2}|1\ 1\rangle$$

$$- \frac{i\hbar^2}{2}|1\ -1\rangle$$

$$= i\hbar^2|1\ 1\rangle$$

$$= i\hbar J_3|1\ 1\rangle.$$

$$[J_1, J_2]|1\ \ 0\rangle = J_1 J_2 |1\ \ 0\rangle - J_2 J_1 |1\ \ 0\rangle$$

$$= \frac{-i\hbar}{\sqrt{2}} J_1 |1\ \ 1\rangle + \frac{i\hbar}{\sqrt{2}} J_1 |1\ \ -1\rangle$$

$$- \frac{\hbar}{\sqrt{2}} J_2 |1\ \ 1\rangle - \frac{\hbar}{\sqrt{2}} J_2 |1\ \ -1\rangle$$

$$= \frac{-i\hbar^2}{2} |1\ \ 0\rangle + \frac{i\hbar^2}{2} |1\ \ 0\rangle - \frac{i\hbar^2}{2} |1\ \ 0\rangle$$

$$+ \frac{i\hbar^2}{2} |1\ \ 0\rangle$$

$$= 0.$$

$$[J_1, J_2]|1\ \ -1\rangle = J_1 J_2 |1\ \ -1\rangle - J_2 J_1 |1\ \ -1\rangle$$

$$= \frac{-i\hbar}{\sqrt{2}} J_1 |1\ \ 0\rangle - \frac{\hbar}{\sqrt{2}} J_2 |1\ \ 0\rangle$$

$$= \frac{-i\hbar^2}{2} |1\ \ 1\rangle - \frac{i\hbar^2}{2} |1\ \ -1\rangle + \frac{i\hbar^2}{2} |1\ \ 1\rangle$$

$$- \frac{i\hbar^2}{2} |1\ \ -1\rangle$$

$$= -i\hbar^2 |1\ \ -1\rangle$$

$$= i\hbar J_3 |1\ \ -1\rangle.$$

$$[J_1, J_3]|1\ \ 1\rangle = J_1 J_3 |1\ \ 1\rangle - J_3 J_1 |1\ \ 1\rangle$$

$$= \hbar J_1 |1\ \ 1\rangle - \frac{\hbar}{\sqrt{2}} J_3 |1\ \ 0\rangle$$

$$= \frac{\hbar^2}{2} |1\ \ 0\rangle$$

$$= -i\hbar J_2 |1\ \ 1\rangle.$$

$$[J_1, J_3]|1\ \ 0\rangle = J_1 J_3 |1\ \ 0\rangle - J_3 J_1 |1\ \ 0\rangle$$

$$= 0 - \frac{\hbar}{\sqrt{2}} J_3 |1\ \ 1\rangle - \frac{\hbar}{\sqrt{2}} J_3 |1\ \ -1\rangle$$

$$= -\frac{\hbar^2}{\sqrt{2}} |1\ \ 1\rangle + \frac{\hbar^2}{\sqrt{2}} |1\ \ -1\rangle$$

$$= -i\hbar J_2 |1\ \ 0\rangle.$$

$$[J_1, J_3]|1 \ -1\rangle = J_1 J_3 |1 \ -1\rangle - J_3 J_1 |1 \ -1\rangle$$

$$= -\hbar J_1 |1 \ -1\rangle - \frac{\hbar}{\sqrt{2}} J_3 |1 \ 0\rangle$$

$$= -\frac{\hbar^2}{\sqrt{2}} |1 \ 0\rangle + 0$$

$$= -i\hbar J_2 |1 \ -1\rangle.$$

$$[J_2, J_3]|1 \ 1\rangle = J_2 J_3 |1 \ 1\rangle - J_3 J_2 |1 \ 1\rangle$$

$$= \hbar J_2 |1 \ 1\rangle - \frac{i\hbar}{\sqrt{2}} J_3 |1 \ 0\rangle$$

$$= \frac{i\hbar^2}{\sqrt{2}} |1 \ 0\rangle + 0$$

$$= i\hbar J_1 |1 \ 1\rangle.$$

$$[J_2, J_3]|1 \ 0\rangle = J_2 J_3 |1 \ 0\rangle - J_3 J_2 |1 \ 0\rangle$$

$$= 0 + \frac{i\hbar}{\sqrt{2}} J_3 |1 \ 1\rangle - \frac{i\hbar}{\sqrt{2}} J_3 |1 \ -1\rangle$$

$$= \frac{i\hbar^2}{\sqrt{2}} |1 \ 1\rangle + \frac{i\hbar^2}{\sqrt{2}} |1 \ -1\rangle$$

$$= i\hbar J_1 |1 \ 0\rangle.$$

$$[J_2, J_3]|1 \ -1\rangle = J_2 J_3 |1 \ -1\rangle - J_3 J_2 |1 \ -1\rangle$$

$$= -\hbar J_2 |1 \ -1\rangle + \frac{i\hbar}{\sqrt{2}} J_3 |1 \ 0\rangle$$

$$= \frac{i\hbar^2}{\sqrt{2}} |1 \ 0\rangle + 0$$

$$= i\hbar J_1 |1 \ -1\rangle.$$

Since the commutation relations hold on the basis vectors, they hold on a general vector.

The matrix corresponding to $J_1$ is

$$J_1 = \begin{pmatrix} \langle 1\ 1|J_1|1\ 1\rangle & \langle 1\ 1|J_1|1\ 0\rangle & \langle 1\ 1|J_1|1\ -1\rangle \\ \langle 1\ 0|J_1|1\ 1\rangle & \langle 1\ 0|J_1|1\ 0\rangle & \langle 1\ 0|J_1|1\ -1\rangle \\ \langle 1\ -1|J_1|1\ 1\rangle & \langle 1\ -1|J_1|1\ 0\rangle & \langle 1\ -1|J_1|1\ -1\rangle \end{pmatrix}$$

$$= \frac{\hbar}{\sqrt{2}} \begin{pmatrix} 0 & 1 & 0 \\ 1 & 0 & 1 \\ 0 & 1 & 0 \end{pmatrix}.$$

Similarly,

$$J_2 = \begin{pmatrix} \langle 1\ 1|J_2|1\ 1\rangle & \langle 1\ 1|J_2|1\ 0\rangle & \langle 1\ 1|J_2|1\ -1\rangle \\ \langle 1\ 0|J_2|1\ 1\rangle & \langle 1\ 0|J_2|1\ 0\rangle & \langle 1\ 0|J_2|1\ -1\rangle \\ \langle 1\ -1|J_2|1\ 1\rangle & \langle 1\ -1|J_2|1\ 0\rangle & \langle 1\ -1|J_2|1\ -1\rangle \end{pmatrix}$$

$$= \frac{i\hbar}{\sqrt{2}} \begin{pmatrix} 0 & -1 & 0 \\ 1 & 0 & -1 \\ 0 & 1 & 0 \end{pmatrix},$$

and

$$J_3 = \begin{pmatrix} \langle 1\ 1|J_3|1\ 1\rangle & \langle 1\ 1|J_3|1\ 0\rangle & \langle 1\ 1|J_3|1\ -1\rangle \\ \langle 1\ 0|J_3|1\ 1\rangle & \langle 1\ 0|J_3|1\ 0\rangle & \langle 1\ 0|J_3|1\ -1\rangle \\ \langle 1\ -1|J_3|1\ 1\rangle & \langle 1\ -1|J_3|1\ 0\rangle & \langle 1\ -1|J_3|1\ -1\rangle \end{pmatrix}$$

$$= \hbar \begin{pmatrix} 1 & 0 & 0 \\ 0 & 0 & 0 \\ 0 & 0 & -1 \end{pmatrix}.$$

Now we test the commutation relations:

$$[J_1, J_2] = \frac{i\hbar^2}{2} \begin{pmatrix} 0 & 1 & 0 \\ 1 & 0 & 1 \\ 0 & 1 & 0 \end{pmatrix} \begin{pmatrix} 0 & -1 & 0 \\ 1 & 0 & -1 \\ 0 & 1 & 0 \end{pmatrix}$$

$$- \frac{i\hbar^2}{2} \begin{pmatrix} 0 & -1 & 0 \\ 1 & 0 & -1 \\ 0 & 1 & 0 \end{pmatrix} \begin{pmatrix} 0 & 1 & 0 \\ 1 & 0 & 1 \\ 0 & 1 & 0 \end{pmatrix}$$

$$= \frac{i\hbar^2}{2} \begin{pmatrix} 1 & 0 & -1 \\ 0 & 0 & 0 \\ 1 & 0 & -1 \end{pmatrix} - \frac{i\hbar^2}{2} \begin{pmatrix} -1 & 0 & -1 \\ 0 & 0 & 0 \\ 1 & 0 & 1 \end{pmatrix}$$

$$= i\hbar^2 \begin{pmatrix} 1 & 0 & 0 \\ 0 & 0 & 0 \\ 0 & 0 & -1 \end{pmatrix} = i\hbar J_3.$$

$$[J_1, J_3] = \frac{\hbar^2}{\sqrt{2}} \begin{pmatrix} 0 & 1 & 0 \\ 1 & 0 & 1 \\ 0 & 1 & 0 \end{pmatrix} \begin{pmatrix} 1 & 0 & 0 \\ 0 & 0 & 0 \\ 0 & 0 & -1 \end{pmatrix}$$

$$- \frac{\hbar^2}{\sqrt{2}} \begin{pmatrix} 1 & 0 & 0 \\ 0 & 0 & 0 \\ 0 & 0 & -1 \end{pmatrix} \begin{pmatrix} 0 & 1 & 0 \\ 1 & 0 & 1 \\ 0 & 1 & 0 \end{pmatrix}$$

$$= \frac{\hbar^2}{\sqrt{2}} \begin{pmatrix} 0 & 0 & 0 \\ 1 & 0 & -1 \\ 0 & 0 & 0 \end{pmatrix} - \frac{\hbar^2}{\sqrt{2}} \begin{pmatrix} 0 & 1 & 0 \\ 0 & 0 & 0 \\ 0 & -1 & 0 \end{pmatrix}$$

$$= \frac{\hbar^2}{\sqrt{2}} \begin{pmatrix} 0 & -1 & 0 \\ 1 & 0 & -1 \\ 0 & 1 & 0 \end{pmatrix} = -i\hbar J_2.$$

$$[J_2, J_3] = \frac{i\hbar^2}{\sqrt{2}} \begin{pmatrix} 0 & -1 & 0 \\ 1 & 0 & -1 \\ 0 & 1 & 0 \end{pmatrix} \begin{pmatrix} 1 & 0 & 0 \\ 0 & 0 & 0 \\ 0 & 0 & -1 \end{pmatrix}$$

$$- \frac{i\hbar^2}{\sqrt{2}} \begin{pmatrix} 1 & 0 & 0 \\ 0 & 0 & 0 \\ 0 & 0 & -1 \end{pmatrix} \begin{pmatrix} 0 & -1 & 0 \\ 1 & 0 & -1 \\ 0 & 1 & 0 \end{pmatrix}$$

$$= \frac{i\hbar^2}{\sqrt{2}} \begin{pmatrix} 0 & 0 & 0 \\ 1 & 0 & 1 \\ 0 & 0 & 0 \end{pmatrix} - \frac{i\hbar^2}{\sqrt{2}} \begin{pmatrix} 0 & -1 & 0 \\ 0 & 0 & 0 \\ 0 & -1 & 0 \end{pmatrix}$$

$$= \frac{i\hbar^2}{\sqrt{2}} \begin{pmatrix} 0 & 1 & 0 \\ 1 & 0 & 1 \\ 0 & 1 & 0 \end{pmatrix} = i\hbar J_1.$$

5. Since $J_1 = \frac{1}{2}(J_+ + J_-)$,

$$J_1^2 = \frac{1}{4}(J_+^2 + J_+J_- + J_-J_+ + J_-^2),$$

and since $J_2 = \frac{-i}{2}(J_+ - J_-)$, then

$$J_2^2 = -\frac{1}{4}(J_+^2 - J_+J_- - J_-J_+ + J_-^2).$$

Also note that

$$\langle j\ m|J_+^2|j\ m\rangle = \langle j\ m|\left(\hbar\sqrt{j(j+1)-m(m+1)}J\right.$$

$$\left.+\ |j\ m+1\rangle\right)$$

$$= \hbar\sqrt{j(j+1)-m(m+1)}\langle j\ m|$$

$$\times\ J_+|j\ m+1\rangle$$

$$= \hbar\sqrt{j(j+1)-m(m+1)}\langle j\ m|$$

$$\times\left(\hbar\sqrt{j(j+1)-(m+1)(m+2)}\right.$$

$$\left.\times\ |j\ m+2\rangle\right)$$

$$= 0.$$

Similarly,

$$\langle j\ m|J_-^2|j\ m\rangle = 0.$$

This means that

$$\langle j\ m|J_1^2|j\ m\rangle = \frac{1}{4}\langle j\ m|J_+^2 + J_+J_- + J_-J_+ + J_-^2|j\ m\rangle$$

$$= \frac{1}{4}\langle j\ m|J_+J_- + J_-J_+|j\ m\rangle$$

$$= \langle j\ m|J_2^2|j\ m\rangle.$$

Now we need

$$\langle j\ m|J_-J_+|j\ m\rangle = \langle j\ m|J_-\left(\hbar\sqrt{j(j+1)-m(m+1)}\right.$$
$$\left.\times|j\ m+1\rangle\right)$$
$$= \hbar\sqrt{j(j+1)-m(m+1)}\langle j\ m|$$
$$\times J_-|j\ m+1\rangle$$
$$= \hbar\sqrt{j(j+1)-m(m+1)}\langle j\ m|$$
$$\times \hbar\sqrt{j(j+1)-(m+1)m}|j\ m\rangle$$
$$= \hbar^2(j(j+1)-m(m+1)),$$

and

$$\langle j\ m|J_+J_-|j\ m\rangle = \hbar\sqrt{j(j+1)-m(m-1)}\langle j\ m|$$
$$\times J_+|j\ m-1\rangle$$
$$= \hbar^2(j(j+1)-m(m-1)),$$

which gives us

$$\langle j\ m|J_1^2|j\ m\rangle = \langle j\ m|J_2^2|j\ m\rangle$$
$$= \frac{\hbar^2}{4}(2j(j+1)-m(m+1)-m(m-1))$$
$$= \frac{\hbar^2}{2}(j(j+1)-m^2).$$

In addition,

$$\langle j\ m|J_1J_2|j\ m\rangle = \langle j\ m|-\tfrac{i}{4}(J_++J_-)(J_+-J_-)|j\ m\rangle$$
$$= -\frac{i}{4}(\langle j\ m|J_+^2|j\ m\rangle - \langle j\ m|J_-^2|j\ m\rangle$$
$$-\langle j\ m|J_+J_-|j\ m\rangle + \langle j\ m|J_-J_+|j\ m\rangle)$$
$$= -\frac{i}{4}(-\hbar^2(j(j+1)-m(m-1))+\hbar^2(j(j+1)$$
$$-m(m+1)))$$
$$= -\frac{i}{4}(-2\hbar^2m) = \frac{im\hbar^2}{2}.$$

6. (a) We have

$$J_1 = \frac{1}{2}(J_+ + J_-)$$

so

$$H = \frac{1}{2I}(J_1^2 + J_3^2) = \frac{1}{8I}(J_+^2 + J_-^2 + J_+J_- + J_-J_+) + \frac{1}{2I}J_3^2.$$

For the final term, we will need

$$\langle 1, m|\, J_3^2\, |1, m\rangle = m\hbar\, \langle 1, m|\, J_3\, |1, m\rangle = m^2\hbar^2.$$

For the other terms, we need

$$J_+^2\, |1, m\rangle = \hbar J_+ \sqrt{(1 - m)(1 + m + 1)}\, |1, m + 1\rangle$$
$$= \hbar\sqrt{(1 - m)(2 + m)}\, J_+\, |1, m + 1\rangle$$
$$= \hbar^2 \sqrt{-m(m + 3)(1 - m)(2 + m)}\, |1, m + 2\rangle$$
$$= \begin{cases} 2\hbar^2\, |1, m + 2\rangle & m = 1 \\ 0 & \text{otherwise.} \end{cases} \tag{*}$$

Thus

$$\langle 1, m|\, J_+^2\, |1, m\rangle = \begin{cases} 2\hbar^2\, \langle 1, m|1, m + 2\rangle & m = -1 \\ 0 & \text{otherwise} \end{cases} = 0.$$

Similarly

$$J_-^2\, |1, m\rangle = \begin{cases} 2\hbar^2\, |1, m - 2\rangle & m = 1 \\ 0 & \text{otherwise} \end{cases} \tag{**}$$

so $\langle 1, m|\, J_-^2\, |1, m\rangle = 0.$

We will also need

$$J_+ J_- \left|1, m\right\rangle = \hbar\sqrt{(1+m)(1-m+1)} J_+ \left|1, m-1\right\rangle$$

$$= \hbar^2 \sqrt{(1+m)(1-m+1)(2-m)(1+m)} \left|1, m\right\rangle$$

$$= \begin{cases} 2\hbar^2 \left|1, m\right\rangle & m = 0, 1 \\ 0 & m = -1, \end{cases} \qquad (\text{***})$$

giving

$$\left\langle 1, m\right| J_+ J_- \left|1, m\right\rangle = \begin{cases} 2\hbar^2 & m = 0, 1 \\ 0 & m = -1. \end{cases}$$

Similarly,

$$\left\langle 1, m\right| J_- J_+ \left|1, m\right\rangle = \begin{cases} 2\hbar^2 & m = -1, 0 \\ 0 & m = 1. \end{cases}$$

Thus

$$\left\langle 1, m\right| J_+ J_- \left|1, m\right\rangle + \left\langle 1, m\right| J_- J_+ \left|1, m\right\rangle = \begin{cases} 2\hbar^2 & m = \pm 1 \\ 4\hbar^2 & m = 0, \end{cases}$$

which we can write more concisely as

$$\left\langle 1, m\right| J_+ J_- \left|1, m\right\rangle + \left\langle 1, m\right| J_- J_+ \left|1, m\right\rangle = 2(2 - m^2)\hbar^2.$$

Using these results, we obtain

$$\left\langle 1, m\right| H \left|1, m\right\rangle = \frac{1}{8I} \left\langle 1, m\right| (J_+^2 + J_-^2 + J_+ J_- + J_- J_+) \left|1, m\right\rangle$$

$$+ \frac{1}{2I} \left\langle 1, m\right| J_3^2 \left|1, m\right\rangle$$

$$= \frac{1}{8I}(0 + 0 + 2(2 - m^2)\hbar^2) + \frac{1}{2I} m^2 \hbar^2$$

$$= \frac{1}{4I}(2 + m^2)\hbar^2$$

as required.

(b) From Eqs. (*), (**) and (***), above,

$$\langle 1, m| H |1, n\rangle = \frac{1}{8I} \langle 1, m| \left( J_+^2 + J_-^2 + J_+ J_- + J_- J_+ \right) |1, n\rangle$$

$$+ \frac{1}{2I} \langle 1, m| J_3^2 |1, n\rangle$$

$$= \frac{1}{8I} \left( 2\hbar^2 \langle 1, m|1, n+2\rangle + 2\hbar^2 \langle 1, m|1, n-2\rangle \right.$$

$$\left. + 2\hbar^2 \langle 1, m|1, n\rangle + 2\hbar^2 \langle 1, m|1, n\rangle \right)$$

$$+ \frac{1}{2I} n^2 \hbar^2 \langle 1, m|1, n\rangle$$

$$= \frac{\hbar^2}{4I} \left( \langle 1, m|1, n+2\rangle + \langle 1, m|1, n-2\rangle \right.$$

$$\left. + 4 \langle 1, m|1, n\rangle \right) + \frac{n^2 \hbar^2}{2I} \langle 1, m|1, n\rangle .$$

Thus, for $m \neq n$, $\langle 1, m| H |1, n\rangle = 0$ unless $m = n+2$ or $m = n-2$, i.e. unless $(m, n) = (1, -1)$ or $(m, n) = (-1, 1)$, in which case

$$\langle 1, m| H |1, n\rangle = \frac{\hbar^2}{4I}.$$

(c) From parts (i) and (ii), $H$ is represented as a matrix in the $\{|1, 1\rangle, |1, 0\rangle, |1, -1\rangle\}$ basis by

$$H = \frac{\hbar^2}{4I} \begin{pmatrix} 3 & 0 & 1 \\ 0 & 2 & 0 \\ 1 & 0 & 3. \end{pmatrix}$$

We can find the eigenvalues and eigenvectors in the usual way (adding a factor of $\hbar^2/4I$ in front of $\lambda$ for convenience):

$$0 = \det \left( \frac{\hbar^2}{4I} \lambda I - H \right) = \frac{\hbar^2}{4I} \begin{vmatrix} \lambda - 3 & 0 & -1 \\ 0 & \lambda - 2 & 0 \\ -1 & 0 & \lambda - 3 \end{vmatrix}$$

$$= \frac{\hbar^2}{4I} (\lambda - 2)^2 (\lambda - 4),$$

so $\lambda = 2$, 2 or 4, and the eigenvalues are $\hbar^2/2I$ (with multiplicity 2) and $\hbar^2/I$.

For $\lambda = 2$, the eigenvectors are given by

$$\begin{pmatrix} 2-3 & 0 & -1 \\ 0 & 2-2 & 0 \\ -1 & 0 & 2-3 \end{pmatrix} \begin{pmatrix} v_1 \\ v_2 \\ v_3 \end{pmatrix} = \begin{pmatrix} -1 & 0 & -1 \\ 0 & 0 & 0 \\ -1 & 0 & -1 \end{pmatrix} \begin{pmatrix} v_1 \\ v_2 \\ v_3 \end{pmatrix} = 0.$$

This gives (as expected) a two-dimensional degenerate subspace of solutions for the eigenvectors corresponding to eigenvalue $\hbar^2/2I$. We can choose any two orthogonal eigenvectors spanning this subspace, e.g.

$$\begin{pmatrix} v_1 \\ v_2 \\ v_3 \end{pmatrix} = \frac{1}{\sqrt{2}} \begin{pmatrix} 1 \\ 0 \\ -1 \end{pmatrix} \quad \text{and} \quad \begin{pmatrix} v_1 \\ v_2 \\ v_3 \end{pmatrix} = \begin{pmatrix} 0 \\ 1 \\ 0 \end{pmatrix}$$

For $\lambda = 4$, we have

$$\begin{pmatrix} 4-3 & 0 & -1 \\ 0 & 4-2 & 0 \\ -1 & 0 & 4-3 \end{pmatrix} \begin{pmatrix} v_1 \\ v_2 \\ v_3 \end{pmatrix} = \begin{pmatrix} 1 & 0 & -1 \\ 0 & 2 & 0 \\ -1 & 0 & 1 \end{pmatrix} \begin{pmatrix} v_1 \\ v_2 \\ v_3 \end{pmatrix} = 0,$$

giving, for the eigenvector corresponding to eigenvalue $\hbar^2/I$,

$$\begin{pmatrix} v_1 \\ v_2 \\ v_3 \end{pmatrix} = \frac{1}{\sqrt{2}} \begin{pmatrix} 1 \\ 0 \\ 1 \end{pmatrix}.$$

7. (a) It is stated in the problem that $J_1$, $J_2$, $J_3$ are self-adjoint. Therefore, one just needs to use the facts (given in class) that products *of the same* self-adjoint operator are self-adjoint and sums and differences of self-adjoint operators are self-adjoint.

   (b)

$$\begin{aligned} J_+ J_- &= (J_1 + iJ_2)(J_1 - iJ_2), \\ &= J_1^2 + J_2^2 - i(J_1 J_2 - J_2 J_1), \\ &= \mathbf{J}^2 - J_3^2 - i[J_1, J_2], \\ &= \mathbf{J}^2 - J_3^2 + \hbar J_3, \end{aligned}$$

$$J_- J_+ = (J_1 - iJ_2)(J_1 + iJ_2),$$
$$= J_1^2 + J_2^2 + i(J_1 J_2 - J_2 J_1),$$
$$= \mathbf{J}^2 - J_3^2 + i[J_1, J_2],$$
$$= \mathbf{J}^2 - J_3^2 - \hbar J_3,$$

Therefore

$$\mathbf{J}^2 = J_+ J_- + J_3^2 - \hbar J_3,$$
$$= J_- J_+ + J_3^2 + \hbar J_3.$$

(c) From the statement of the problem, we have:

$$\mathbf{J}^2 \mid jm\rangle = \hbar^2 j(j+1) \mid jm\rangle, \quad J_3 \mid jm\rangle = \hbar m \mid jm\rangle.$$

From which it follows that:

$$J_3^2 \mid jm\rangle = \hbar^2 m^2 \mid jm\rangle,$$

and therefore

$$H \mid jm\rangle = (\mathbf{J}^2 - J_3^2) \mid jm\rangle = \hbar^2 (j(j+1) - m^2) \mid jm\rangle,$$

and the eigenvalues of $H$ are

$$\hbar^2 (j(j+1) - m^2).$$

(d) Yes. This follows from the result of part (c), and the student should say that $\mid jm\rangle$ are eigenvectors of $H$ and $\mathbf{J}^2$, as was shown in part (c).

(e)

$$\begin{pmatrix} \langle \tfrac{1}{2}, \tfrac{1}{2} \mid H \mid \tfrac{1}{2}, \tfrac{1}{2} \rangle & \langle \tfrac{1}{2}, \tfrac{1}{2} \mid H \mid \tfrac{1}{2}, -\tfrac{1}{2} \rangle \\ \langle \tfrac{1}{2}, -\tfrac{1}{2} \mid H \mid \tfrac{1}{2}, \tfrac{1}{2} \rangle & \langle \tfrac{1}{2}, -\tfrac{1}{2} \mid H \mid \tfrac{1}{2}, -\tfrac{1}{2} \rangle \end{pmatrix}$$

Using result from part (d) and orthonormality of the eigenstates, this expression becomes:

$$\frac{\hbar^2}{2} \begin{pmatrix} 1 & 0 \\ 0 & 1 \end{pmatrix}$$

8. (a) Using the definition of the cross product and the commutation relations for the $J_i$, $i = 1, 2, 3$, gives:

$$\mathbf{J} \times \mathbf{J} = (J_2 J_3 - J_3 J_2, J_3 J_1 - J_1 J_3, J_1 J_2 - J_2 J_1),$$
$$= ([J_2, J_3], [J_3, J_1], [J_1, J_2]),$$
$$= i\hbar (J_1, J_2, J_3),$$
$$= i\hbar \mathbf{J}.$$

(b)

$$J_+ J_- = (J_1 + iJ_2)(J_1 - iJ_2),$$
$$= J_1^2 + J_2^2 - i(J_1 J_2 - J_2 J_1),$$
$$= \mathbf{J}^2 - J_3^2 - i[J_1, J_2],$$
$$= \mathbf{J}^2 - J_3^2 + \hbar J_3,$$

$$J_- J_+ = (J_1 - iJ_2)(J_1 + iJ_2),$$
$$= J_1^2 + J_2^2 + i(J_1 J_2 - J_2 J_1),$$
$$= \mathbf{J}^2 - J_3^2 + i[J_1, J_2],$$
$$= \mathbf{J}^2 - J_3^2 - \hbar J_3,$$

Therefore

$$\mathbf{J}^2 = J_+ J_- + J_3^2 - \hbar J_3,$$
$$= J_- J_+ + J_3^2 + \hbar J_3.$$

(c) From the statement of the problem, we have:

$$\mathbf{J}^2 \mid jm \rangle = \hbar^2 j(j+1) \mid jm \rangle, \quad J_3 \mid jm \rangle = \hbar m \mid jm \rangle.$$

From which it follows that:

$$J_3^2 \mid jm \rangle = \hbar^2 m^2 \mid jm \rangle,$$

and therefore

$$H \mid jm \rangle = \left( \frac{1}{2I_1} \mathbf{J}^2 - \left( \frac{1}{2I_1} - \frac{1}{2I_3} \right) J_3^2 \right) \mid jm \rangle$$
$$= \left( \frac{\hbar^2 (j(j+1))}{2I_1} - \hbar^2 m^2 \left( \frac{1}{2I_1} - \frac{1}{2I_3} \right) \right) \mid jm \rangle,$$

and the eigenvalues of $H$ are

$$\frac{\hbar^2 (j(j+1))}{2I_1} - \hbar^2 m^2 \left( \frac{1}{2I_1} - \frac{1}{2I_3} \right).$$

(d) The matrix representation has the form:

$$\begin{pmatrix} \langle \tfrac{1}{2}, \tfrac{1}{2} \mid H \mid \tfrac{1}{2}, \tfrac{1}{2} \rangle & \langle \tfrac{1}{2}, \tfrac{1}{2} \mid H \mid \tfrac{1}{2}, -\tfrac{1}{2} \rangle \\ \langle \tfrac{1}{2}, -\tfrac{1}{2} \mid H \mid \tfrac{1}{2}, \tfrac{1}{2} \rangle & \langle \tfrac{1}{2}, -\tfrac{1}{2} \mid H \mid \tfrac{1}{2}, -\tfrac{1}{2} \rangle \end{pmatrix}$$

Using result from part (d), note that the eigenvalue for $|\frac{1}{2}, \frac{1}{2}\rangle$ is the same as the eigenvalue for $.\ |\frac{1}{2}, -\frac{1}{2}\rangle$, and these eigenvalues are given by:

$$\frac{\hbar^2}{4I_1} + \frac{\hbar^2}{8I_3} \equiv \lambda.$$

Using this result, and orthogonality of the eigenstates, the matrix representation becomes:

$$\begin{pmatrix} \lambda & 0 \\ 0 & \lambda \end{pmatrix}$$

(e) $\langle \frac{1}{2}, -\frac{1}{2} | J_3^2 | \frac{1}{2}, -\frac{1}{2}\rangle = \hbar^2 m^2.$

9. (a) It is easy to verify that:

$$S^2 = \begin{pmatrix} 3 & 0 \\ 0 & 3 \end{pmatrix},$$

Therefore we have:

$$S^2 X - X S^2 = \begin{pmatrix} 3 & 0 \\ 0 & 3 \end{pmatrix}\begin{pmatrix} 0 & 1 \\ 1 & 0 \end{pmatrix} - \begin{pmatrix} 0 & 1 \\ 1 & 0 \end{pmatrix}\begin{pmatrix} 3 & 0 \\ 0 & 3 \end{pmatrix},$$

$$= \begin{pmatrix} 0 & 3 \\ 3 & 0 \end{pmatrix} - \begin{pmatrix} 0 & 3 \\ 3 & 0 \end{pmatrix} = \begin{pmatrix} 0 & 0 \\ 0 & 0 \end{pmatrix}.$$

$$S^2 Y - Y S^2 = \begin{pmatrix} 3 & 0 \\ 0 & 3 \end{pmatrix}\begin{pmatrix} 0 & -i \\ i & 0 \end{pmatrix} - \begin{pmatrix} 0 & -i \\ i & 0 \end{pmatrix}\begin{pmatrix} 3 & 0 \\ 0 & 3 \end{pmatrix},$$

$$= \begin{pmatrix} 0 & -3i \\ 3i & 0 \end{pmatrix} - \begin{pmatrix} 0 & -3i \\ 3i & 0 \end{pmatrix} = \begin{pmatrix} 0 & 0 \\ 0 & 0 \end{pmatrix}.$$

$$S^2 Y - Y S^2 = \begin{pmatrix} 3 & 0 \\ 0 & 3 \end{pmatrix}\begin{pmatrix} 0 & -i \\ i & 0 \end{pmatrix} - \begin{pmatrix} 0 & -i \\ i & 0 \end{pmatrix}\begin{pmatrix} 3 & 0 \\ 0 & 3 \end{pmatrix},$$

$$= \begin{pmatrix} 0 & -3i \\ 3i & 0 \end{pmatrix} - \begin{pmatrix} 0 & -3i \\ 3i & 0 \end{pmatrix} = \begin{pmatrix} 0 & 0 \\ 0 & 0 \end{pmatrix}.$$

$$S^2 Z - Z S^2 = \begin{pmatrix} 3 & 0 \\ 0 & 3 \end{pmatrix}\begin{pmatrix} 1 & 0 \\ 0 & -1 \end{pmatrix} - \begin{pmatrix} 1 & 0 \\ 0 & -1 \end{pmatrix}\begin{pmatrix} 3 & 0 \\ 0 & 3 \end{pmatrix},$$

$$= \begin{pmatrix} 3 & 0 \\ 0 & -3 \end{pmatrix} - \begin{pmatrix} 3 & 0 \\ 0 & -3 \end{pmatrix} = \begin{pmatrix} 0 & 0 \\ 0 & 0 \end{pmatrix}.$$

(b) It is straightforward to verify that

$$X^\dagger = \overline{X}^T = X, \quad Y^\dagger = \overline{Y}^T = Y, \quad Z^\dagger = \overline{Z}^T = Z.$$

and therefore $X$, $Y$, and $Z$ are self-adjoint.
However,

$$\left(S^+\right)^\dagger = X^\dagger - iY^\dagger = X - iY \neq S^+,$$

and

$$\left(S^-\right)^\dagger = X^\dagger + iY^\dagger = X - iY \neq S^-.$$

Therefore $S^+$ and $S^-$ are not self-adjoint.

(c)

$$S^+ \begin{pmatrix} 0 \\ 1 \end{pmatrix} = \begin{pmatrix} 0 & 2 \\ 0 & 0 \end{pmatrix} \begin{pmatrix} 0 \\ 1 \end{pmatrix} = \begin{pmatrix} 2 \\ 0 \end{pmatrix},$$

which is an eigenvector if $Z$. Similarly,

$$S^- \begin{pmatrix} 1 \\ 0 \end{pmatrix} = \begin{pmatrix} 0 & 0 \\ 2 & 0 \end{pmatrix} \begin{pmatrix} 1 \\ 0 \end{pmatrix} = \begin{pmatrix} 0 \\ 2 \end{pmatrix},$$

which is also an eigenvector of $Z$.

(d) $S^2$ and $Y$ commute. Therefore we know by the theorem related to simultaneous measurability given in class that a set of vectors that are simultaneously eigenvectors for $S^2$ and $Y$ exist. It is a simple calculation to check that the two eigenvectors of $S^2$ are also eigenvectors of $Y$, and that the two eigenvectors of $Y$ are also eigenvectors of $S^2$.


## Chapter 5

1. (a) Since $J(\theta, \phi) = \sin\theta \, \cos\phi \, J_x + \sin\theta \, \sin\phi \, J_y + \cos\theta \, J_z$, we can write the matrix of $J(\theta, \phi)$ (using the Pauli spin matrices) as

$$J(\theta, \phi) = \begin{pmatrix} \frac{\hbar}{2}\cos\theta & \frac{\hbar}{2}\sin\theta(\cos\phi - i\sin\phi) \\ \frac{\hbar}{2}\sin\theta(\cos\phi + i\sin\phi) & -\frac{\hbar}{2}\cos\theta \end{pmatrix}$$

$$= \begin{pmatrix} \frac{\hbar}{2}\cos\theta & \frac{\hbar}{2}\sin\theta \, e^{-i\phi} \\ \frac{\hbar}{2}\sin\theta \, e^{i\phi} & -\frac{\hbar}{2}\cos\theta \end{pmatrix}.$$

To calculate the eigenvalues:

$$\begin{vmatrix} \frac{\hbar}{2}\cos\theta - \lambda & \frac{\hbar}{2}\sin\theta\, e^{-i\phi} \\ \frac{\hbar}{2}\sin\theta\, e^{i\phi} & -\frac{\hbar}{2}\cos\theta - \lambda \end{vmatrix} = 0$$

$$\left(\frac{\hbar}{2}\cos\theta - \lambda\right)\left(-\frac{\hbar}{2}\cos\theta - \lambda\right) - \frac{\hbar^2}{4}\sin^2\theta = 0$$

$$-\frac{\hbar^2}{4}\cos^2\theta + \lambda^2 - \frac{\hbar^2}{4}\sin^2\theta = 0$$

$$\lambda^2 - \frac{\hbar^2}{4} = 0 \;\Rightarrow\; \lambda = \pm\frac{\hbar}{2}.$$

If $\lambda = \frac{\hbar}{2}$,

$$\begin{pmatrix} \frac{\hbar}{2}(\cos\theta - 1) & \frac{\hbar}{2}\sin\theta\, e^{-i\phi} \\ \frac{\hbar}{2}\sin\theta\, e^{i\phi} & -\frac{\hbar}{2}(\cos\theta + 1) \end{pmatrix} \begin{pmatrix} a \\ b \end{pmatrix} = \begin{pmatrix} 0 \\ 0 \end{pmatrix}$$

$$\Rightarrow \; (\cos\theta - 1)a + \sin\theta\, e^{-i\phi}b = 0.$$

The choices $a = \cos\frac{\theta}{2}$ and $b = \sin\frac{\theta}{2}\, e^{i\phi}$ satisfies this condition:

$$(\cos\theta - 1)\cos\frac{\theta}{2} + \sin\theta\, e^{-i\phi}e^{i\phi}\sin\frac{\theta}{2}$$

$$= \left(2\cos^2\frac{\theta}{2} - 2\right)\cos\frac{\theta}{2} + 2\sin\frac{\theta}{2}\,\cos\frac{\theta}{2}\,\sin\frac{\theta}{2}$$

$$= -2\cos\frac{\theta}{2} + 2\cos\frac{\theta}{2}\left(\cos^2\frac{\theta}{2} + \sin^2\frac{\theta}{2}\right) = 0,$$

so

$$|\theta, \phi\rangle = \cos\frac{\theta}{2}|\tfrac{1}{2}\;\tfrac{1}{2}\rangle + e^{i\phi}\sin\frac{\theta}{2}|\tfrac{1}{2}\; -\tfrac{1}{2}\rangle$$

is an eigenvector.

If $\lambda = -\frac{\hbar}{2}$,

$$\begin{pmatrix} \frac{\hbar}{2}(\cos\theta + 1) & \frac{\hbar}{2}\sin\theta\, e^{-i\phi} \\ \frac{\hbar}{2}\sin\theta\, e^{i\phi} & -\frac{\hbar}{2}(\cos\theta - 1) \end{pmatrix} \begin{pmatrix} a \\ b \end{pmatrix} = \begin{pmatrix} 0 \\ 0 \end{pmatrix}$$

$$\Rightarrow \; (\cos\theta + 1)a + \sin\theta\, e^{-i\phi}b = 0.$$

The choices $a = \sin\frac{\theta}{2}$ and $b = -\cos\frac{\theta}{2}\, e^{i\phi}$ satisfies this condition:

$$(\cos\theta + 1)\sin\frac{\theta}{2} + \sin\theta\, e^{-i\phi}e^{i\phi}\cos\frac{\theta}{2}$$

$$= 2\cos^2\frac{\theta}{2}\,\sin\frac{\theta}{2} - 2\cos^2\frac{\theta}{2}\,\sin\frac{\theta}{2} = 0,$$

so

$$|\pi - \theta, \phi + \pi\rangle = \sin\tfrac{\theta}{2}|\tfrac{1}{2}\ \tfrac{1}{2}\rangle - e^{i\phi}\cos\tfrac{\theta}{2}|\tfrac{1}{2}\ -\tfrac{1}{2}\rangle$$

is an eigenvector.

(b) The vectors above are normalised because

$$|\cos\tfrac{\theta}{2}|^2 + |e^{i\phi}\sin\tfrac{\theta}{2}|^2 = \cos^2\tfrac{\theta}{2} + \sin^2\tfrac{\theta}{2} = 1$$

and

$$|\sin\tfrac{\theta}{2}|^2 + |-e^{i\phi}\cos\tfrac{\theta}{2}|^2 = \cos^2\tfrac{\theta}{2} + \sin^2\tfrac{\theta}{2} = 1.$$

They are orthogonal because

$$\langle\pi - \theta, \phi + \pi|\theta, \phi\rangle$$
$$= (\sin\tfrac{\theta}{2}\langle\tfrac{1}{2}\ \tfrac{1}{2}| - e^{-i\phi}\cos\tfrac{\theta}{2}\langle\tfrac{1}{2}\ -\tfrac{1}{2}|)$$
$$\times (\cos\tfrac{\theta}{2}|\tfrac{1}{2}\ \tfrac{1}{2}\rangle + e^{i\phi}\sin\tfrac{\theta}{2}|\tfrac{1}{2}\ -\tfrac{1}{2}\rangle)$$
$$= \sin\tfrac{\theta}{2}\cos\tfrac{\theta}{2} - \cos\tfrac{\theta}{2}\sin\tfrac{\theta}{2} = 0,$$

due to the othonormality of $|\tfrac{1}{2}\ \tfrac{1}{2}\rangle$ and $|\tfrac{1}{2}\ -\tfrac{1}{2}\rangle$.

(c) Since $|\theta, \phi\rangle$ and $|\pi - \theta, \phi + \pi\rangle$ are eigenvectors of $J(\theta, \phi)$ with eigenvalues $\tfrac{\hbar}{2}$ and $-\tfrac{\hbar}{2}$ respectively, this is just the usual expression of an operator in terms of the projectors onto the spaces spanned by the eigenvectors.

2. Since the state $|\tfrac{1}{2}\ \tfrac{1}{2}\rangle$ is normalised, the probability that $\hbar/2$ is found when $J(\theta, \phi)$ is measured is

$$|\langle\tfrac{1}{2}\ \tfrac{1}{2}|\theta, \phi\rangle|^2 = |\langle\tfrac{1}{2}\ \tfrac{1}{2}|(\cos\tfrac{\theta}{2}|\tfrac{1}{2}\ \tfrac{1}{2}\rangle + e^{i\phi}$$
$$\times \sin\tfrac{\theta}{2}|\tfrac{1}{2}\ -\tfrac{1}{2}\rangle)|^2$$
$$= \cos^2\tfrac{\theta}{2}.$$

3. (a) As described in the lectures, if operators $A$, $B$ and $C$ are related by $[A, B] = iC$, we consider the operator $(A - itB)$. Its adjoint is $(A + itB)$ so we have

$$(A - itB)^\dagger(A - itB) = (A + itB)(A - itB)$$
$$= A^2 - it(AB - BA) + t^2 B^2.$$

Assume $\psi$ is normalized. Thus

$$E_{|\psi\rangle}(A^2) + tE_{|\psi\rangle}(C) + t^2 E_{|\psi\rangle}(B^2)$$
$$= E_{|\psi\rangle}(A^2 + tC + t^2 B^2)$$
$$= E_{|\psi\rangle}((A - itB)^\dagger (A - itB))$$
$$= \langle\psi|((A - itB)^\dagger (A - itB)|\psi\rangle$$
$$= ||(A - itB)|\psi\rangle||^2. \tag{5.1}$$

This implies that $E_{|\psi\rangle}(A^2) + tE_{|\psi\rangle}(C) + t^2 E_{|\psi\rangle}(B^2) \geq 0$. Since this quadratic expression is non-negative, its discriminant $E_{|\psi\rangle}(C)^2 - 4E_{|\psi\rangle}(A^2)E_{|\psi\rangle}(B^2)$ must be non-positive. In other words

$$E_{|\psi\rangle}(A^2)E_{|\psi\rangle}(B^2) \geq \frac{1}{4}E_{|\psi\rangle}(C)^2.$$

The equality holds when the quadratic above has a repeated real root $t$. Thus, from (5.1), the equality holds if and only if there exists a real number $t$ such that $(A - itB)|\psi\rangle = 0$.

To relate this to the angular momentum operators, let $A = J_x - E_{|\psi\rangle}(J_x)$ and $B = J_y - E_{|\psi\rangle}(J_y)$. Then we have

$$[A, B] = [J_x, J_y] - E_{|\psi\rangle}(J_y)[J_x, 1] - E_{|\psi\rangle}(J_x)[1, J_y]$$
$$+ E_{|\psi\rangle}(J_x)E_{|\psi\rangle}(J_y)[1, 1]$$
$$= [J_x, J_y] = i\hbar J_z,$$

so

$$E_{|\psi\rangle}((J_x - E_{|\psi\rangle}(J_x))^2)$$
$$\times E_{|\psi\rangle}((J_y - E_{|\psi\rangle}(J_y))^2) \geq \frac{1}{4}E_{|\psi\rangle}(\hbar J_z)^2$$
$$\Delta_{|\psi\rangle}(J_x)^2 \Delta_{|\psi\rangle}(J_y)^2 \geq \frac{1}{4}E_{|\psi\rangle}(\hbar J_z)^2.$$

and

$$\Delta_{|\psi\rangle}(J_x)\Delta_{|\psi\rangle}(J_y) \geq \frac{\hbar}{2}|E_{|\psi\rangle}(J_z)|.$$

The equality above holds when there exists a real number $t$ such that $((J_x - E_{|\psi\rangle}(J_x)) - it(J_y - E_{|\psi\rangle}(J_y)))|\psi\rangle = 0$.

(b) The vector $|\tfrac{1}{2}\ \tfrac{1}{2}\rangle$ is normalised, so

$$\Delta_{|\frac{1}{2}\ \frac{1}{2}\rangle}(J_x) = \left(E_{|\frac{1}{2}\ \frac{1}{2}\rangle}(J_x^2) - E_{|\frac{1}{2}\ \frac{1}{2}\rangle}(J_x)^2\right)^{1/2}$$

$$= \left(\langle\tfrac{1}{2}\ \tfrac{1}{2}|J_x^2|\tfrac{1}{2}\ \tfrac{1}{2}\rangle - \langle\tfrac{1}{2}\ \tfrac{1}{2}|J_x|\tfrac{1}{2}\ \tfrac{1}{2}\rangle^2\right)^{1/2}$$

$$= \left(\langle\tfrac{1}{2}\ \tfrac{1}{2}|(\tfrac{1}{2}(J_+ + J_-))^2|\tfrac{1}{2}\ \tfrac{1}{2}\rangle\right.$$

$$\left. -\langle\tfrac{1}{2}\ \tfrac{1}{2}|\tfrac{1}{2}(J_+ + J_-)|\tfrac{1}{2}\ \tfrac{1}{2}\rangle^2\right)^{1/2}$$

$$= \left(\langle\tfrac{1}{2}\ \tfrac{1}{2}|\tfrac{1}{2}(J_+ + J_-))\tfrac{1}{2}\hbar|\tfrac{1}{2}\right.$$

$$\left. -\tfrac{1}{2}\rangle - \langle\tfrac{1}{2}\ \tfrac{1}{2}|\tfrac{1}{2}\hbar|\tfrac{1}{2}\ -\tfrac{1}{2}\rangle^2\right)^{1/2}$$

$$= \left(\tfrac{\hbar}{4}\langle\tfrac{1}{2}\ \tfrac{1}{2}|\hbar|\tfrac{1}{2}\ \tfrac{1}{2}\rangle - 0\right)^{1/2}$$

$$= \frac{\hbar}{2}$$

and

$$\Delta_{|\frac{1}{2}\ \frac{1}{2}\rangle}(J_y) = \left(\langle\tfrac{1}{2}\ \tfrac{1}{2}|J_y^2|\tfrac{1}{2}\ \tfrac{1}{2}\rangle - \langle\tfrac{1}{2}\ \tfrac{1}{2}|J_y|\tfrac{1}{2}\ \tfrac{1}{2}\rangle^2\right)^{1/2}$$

$$= \left(\langle\tfrac{1}{2}\ \tfrac{1}{2}|(\tfrac{-i}{2}(J_+ - J_-))^2|\tfrac{1}{2}\ \tfrac{1}{2}\rangle\right.$$

$$\left. -\langle\tfrac{1}{2}\ \tfrac{1}{2}|\tfrac{-i}{2}(J_+ - J_-)|\tfrac{1}{2}\ \tfrac{1}{2}\rangle^2\right)^{1/2}$$

$$= \left(\tfrac{-1}{4}\langle\tfrac{1}{2}\ \tfrac{1}{2}|(J_+ - J_-)(-\hbar)|\tfrac{1}{2}\ -\tfrac{1}{2}\rangle\right.$$

$$\left. +\frac{i}{2}\langle\tfrac{1}{2}\ \tfrac{1}{2}|-\hbar|\tfrac{1}{2}\ -\tfrac{1}{2}\rangle^2\right)^{1/2}$$

$$= \left(\frac{\hbar^2}{4} - 0\right)^{1/2}$$

$$= \frac{\hbar}{2}.$$

In addition,

$$|E_{|\frac{1}{2}\ \frac{1}{2}\rangle}(J_z)| = |\langle\tfrac{1}{2}\ \tfrac{1}{2}|J_z|\tfrac{1}{2}\ \tfrac{1}{2}\rangle|$$

$$= |\langle\tfrac{1}{2}\ \tfrac{1}{2}|\tfrac{\hbar}{2}|\tfrac{1}{2}\ \tfrac{1}{2}\rangle| = \frac{\hbar}{2}.$$

So, the uncertainty relation reads

$$\frac{\hbar}{2}\cdot\frac{\hbar}{2} \geq \frac{\hbar^2}{4},$$

so this is a case of equality.

Check: is there a real $t$ such that

$$\left(J_x - E_{|\frac{1}{2}\ \frac{1}{2}\rangle}(J_x) - itJ_y + itE_{|\frac{1}{2}\ \frac{1}{2}\rangle}(J_y)\right)|\tfrac{1}{2}\ \tfrac{1}{2}\rangle = 0?$$

We already know from the calculation above that $E_{|\frac{1}{2}\ \frac{1}{2}\rangle}(J_x) = E_{|\frac{1}{2}\ \frac{1}{2}\rangle}(J_y) = 0$, so consider

$$J_x|\tfrac{1}{2}\ \tfrac{1}{2}\rangle - itJ_y|\tfrac{1}{2}\ \tfrac{1}{2}\rangle = \frac{\hbar}{2}|\tfrac{1}{2}\ -\tfrac{1}{2}\rangle - it\left(\frac{i\hbar}{2}|\tfrac{1}{2}\ -\tfrac{1}{2}\rangle\right)$$

$$= \frac{\hbar}{2}|\tfrac{1}{2}\ -\tfrac{1}{2}\rangle + \frac{\hbar}{2}t|\tfrac{1}{2}\ -\tfrac{1}{2}\rangle.$$

This equals zero if $t = -1$. So, we do indeed expect to achieve the lower bound in the uncertainty relation.

(c) You might recognise that this is an eigenstate of $J_x$ and so you know the dispersion is zero. However, you can calculate it as follows:

$$\Delta_{|\psi\rangle}(J_x) = \left(\left(\frac{1}{\sqrt{2}}\langle\tfrac{1}{2}\ \tfrac{1}{2}| + \frac{1}{\sqrt{2}}\langle\tfrac{1}{2}\ -\tfrac{1}{2}|\right)\right.$$

$$\times\ J_x^2\left(\frac{1}{\sqrt{2}}|\tfrac{1}{2}\ \tfrac{1}{2}\rangle + \frac{1}{\sqrt{2}}|\tfrac{1}{2}\ -\tfrac{1}{2}\rangle\right)$$

$$-\left(\left(\frac{1}{\sqrt{2}}\langle\tfrac{1}{2}\ \tfrac{1}{2}| + \frac{1}{\sqrt{2}}\langle\tfrac{1}{2}\ -\tfrac{1}{2}|\right)\right.$$

$$\left.\left.\times\ J_x\left(\frac{1}{\sqrt{2}}|\tfrac{1}{2}\ \tfrac{1}{2}\rangle + \frac{1}{\sqrt{2}}|\tfrac{1}{2}\ -\tfrac{1}{2}\rangle\right)\right)^2\right)^{1/2}$$

$$= \left(\left(\frac{1}{\sqrt{2}}\langle\tfrac{1}{2}\ \tfrac{1}{2}| + \frac{1}{\sqrt{2}}\langle\tfrac{1}{2}\ -\tfrac{1}{2}|\right)\right.$$

$$\times\ J_x\left(\frac{\hbar}{2\sqrt{2}}|\tfrac{1}{2}\ -\tfrac{1}{2}\rangle + \frac{\hbar}{2\sqrt{2}}|\tfrac{1}{2}\ \tfrac{1}{2}\rangle\right)$$

$$-\left(\left(\frac{1}{\sqrt{2}}\langle\tfrac{1}{2}\ \tfrac{1}{2}| + \frac{1}{\sqrt{2}}\langle\tfrac{1}{2}\ -\tfrac{1}{2}|\right)\right.$$

$$\left.\left.\left(\frac{\hbar}{2\sqrt{2}}|\tfrac{1}{2}\ -\tfrac{1}{2}\rangle + \frac{\hbar}{2\sqrt{2}}|\tfrac{1}{2}\ \tfrac{1}{2}\rangle\right)\right)^2\right)^{1/2}$$

$$= \left(\left(\frac{1}{\sqrt{2}}\langle\tfrac{1}{2}\ \tfrac{1}{2}| + \frac{1}{\sqrt{2}}\langle\tfrac{1}{2}\ -\tfrac{1}{2}|\right)\frac{\hbar}{2\sqrt{2}}\right.$$

$$\left.\left(\tfrac{\hbar}{2}|\tfrac{1}{2}\ \tfrac{1}{2}\rangle + \tfrac{\hbar}{2}|\tfrac{1}{2}\ -\tfrac{1}{2}\rangle\right) - \left(\tfrac{\hbar}{4} + \tfrac{\hbar}{4}\right)^2\right)^{1/2}$$

$$= \left(\left(\frac{\hbar^2}{8} + \frac{\hbar^2}{8}\right) - \frac{\hbar^2}{4}\right)^{1/2} = 0.$$

We also see that

$$E_{|\psi\rangle}(J_z) = \left(\tfrac{1}{\sqrt{2}}\right)^2 \cdot \tfrac{\hbar}{2} + \left(\tfrac{1}{\sqrt{2}}\right)^2 \cdot \left(-\tfrac{\hbar}{2}\right) = 0,$$

so the uncertainty relation is satisfied.

4. (a)

$$(X \otimes \mathbb{I})(X \otimes \mathbb{I})^{\dagger} = (X \otimes \mathbb{I})\left(X^{\dagger} \otimes \mathbb{I}^{\dagger}\right),$$
$$= (X \otimes \mathbb{I})(X \otimes \mathbb{I}),$$
$$= X^2 \otimes \mathbb{I},$$
$$= \mathbb{I} \otimes \mathbb{I}.$$

(b)

$$|\psi\rangle = \frac{1}{\sqrt{2}}\left(|1\rangle |2\rangle - |2\rangle |1\rangle\right) \tag{5.2}$$

$$|\psi_X\rangle = (X \otimes \mathbb{I})|\psi\rangle,$$
$$= \frac{1}{\sqrt{2}}\left(|2\rangle |2\rangle - |1\rangle |1\rangle\right). \tag{5.3}$$

(c)

$$|\psi_Y\rangle = (Y \otimes \mathbb{I})|\psi\rangle,$$
$$= \frac{i}{\sqrt{2}}\left(|2\rangle |2\rangle + |1\rangle |1\rangle\right). \tag{5.4}$$

$$|\psi_Z\rangle = (Z \otimes \mathbb{I})|\psi\rangle,$$
$$= \frac{1}{\sqrt{2}}\left(|1\rangle |2\rangle + |2\rangle |1\rangle\right). \tag{5.5}$$

$$\langle\psi|\psi\rangle, \ \langle\psi|\psi_X\rangle, \ \langle\psi|\psi_Y\rangle, \ \langle\psi|\psi_Z\rangle \tag{5.6}$$

$$\langle\psi|\psi\rangle = \frac{1}{2}\left(\langle 1|\langle 2| - \langle 2|\langle 1|\right)\left(|1\rangle |2\rangle - |2\rangle |1\rangle\right),$$
$$= \frac{1}{2}\left(\left(\langle 1|\langle 2|\right)\left(|1\rangle |2\rangle\right) - \left(\langle 1|\langle 2|\right)\left(|2\rangle |1\rangle\right)\right.$$
$$\left. - \left(\langle 2|\langle 1|\right)\left(|1\rangle |2\rangle\right) + \left(\langle 2|\langle 1|\right)\left(|2\rangle |1\rangle\right)\right),$$
$$= \frac{1}{2}\left(\langle 1|1\rangle\langle 2|2\rangle - \langle 1|2\rangle\langle 2|1\rangle\right.$$
$$\left. - \langle 2|1\rangle\langle 1|2\rangle + \langle 2|2\rangle\langle 1|1\rangle\right),$$
$$= \frac{1}{2}\left(1 - 0 - 0 + 1\right) = 1 \tag{5.7}$$

$$\langle \psi \mid \psi_X \rangle = \frac{1}{2} \left( \langle 1 \mid \langle 2 \mid - \langle 2 \mid \langle 1 \mid \right) \left( \mid 2 \rangle \mid 2 \rangle - \mid 1 \rangle \mid 1 \rangle \right),$$

$$= \frac{1}{2} \left( \left( \langle 1 \mid \langle 2 \mid \right) \left( \mid 2 \rangle \mid 2 \rangle \right) - \left( \langle 1 \mid \langle 2 \mid \right) \left( \mid 1 \rangle \mid 1 \rangle \right) \right.$$

$$\left. - \left( \langle 2 \mid \langle 1 \mid \right) \left( \mid 2 \rangle \mid 2 \rangle \right) + \left( \langle 2 \mid \langle 1 \mid \right) \left( \mid 1 \rangle \mid 1 \rangle \right) \right),$$

$$= \frac{1}{2} \left( \langle 1 \mid 2 \rangle \langle 2 \mid 2 \rangle - \langle 1 \mid 1 \rangle \langle 2 \mid 1 \rangle \right.$$

$$\left. - \langle 2 \mid 2 \rangle \langle 1 \mid 2 \rangle + \langle 2 \mid 1 \rangle \langle 1 \mid 1 \rangle \right),$$

$$= \frac{1}{2} \left( 0 - 0 - 0 + 0 \right) = 0 \tag{5.8}$$

$$\langle \psi \mid \psi_Y \rangle = \frac{i}{2} \left( \langle 1 \mid \langle 2 \mid - \langle 2 \mid \langle 1 \mid \right) \left( \mid 2 \rangle \mid 2 \rangle + \mid 1 \rangle \mid 1 \rangle \right),$$

$$= \frac{1}{2} \left( \left( \langle 1 \mid \langle 2 \mid \right) \left( \mid 2 \rangle \mid 2 \rangle \right) + \left( \langle 1 \mid \langle 2 \mid \right) \left( \mid 1 \rangle \mid 1 \rangle \right) \right.$$

$$\left. - \left( \langle 2 \mid \langle 1 \mid \right) \left( \mid 2 \rangle \mid 2 \rangle \right) - \left( \langle 2 \mid \langle 1 \mid \right) \left( \mid 1 \rangle \mid 1 \rangle \right) \right),$$

$$= \frac{1}{2} \left( \langle 1 \mid 2 \rangle \langle 2 \mid 2 \rangle + \langle 1 \mid 1 \rangle \langle 2 \mid 1 \rangle - \langle 2 \mid 2 \rangle \langle 1 \mid 2 \rangle \right.$$

$$\left. - \langle 2 \mid 1 \rangle \langle 1 \mid 1 \rangle \right),$$

$$= \frac{1}{2} \left( 0 + 0 - 0 - 0 \right) = 0 \tag{5.9}$$

$$\langle \psi \mid \psi_Z \rangle = \frac{1}{2} \left( \langle 1 \mid \langle 2 \mid - \langle 2 \mid \langle 1 \mid \right) \left( \mid 1 \rangle \mid 2 \rangle + \mid 2 \rangle \mid 1 \rangle \right),$$

$$= \frac{1}{2} \left( \left( \langle 1 \mid \langle 2 \mid \right) \left( \mid 2 \rangle \mid 2 \rangle \right) + \left( \langle 1 \mid \langle 2 \mid \right) \left( \mid 2 \rangle \mid 1 \rangle \right) \right.$$

$$\left. - \left( \langle 2 \mid \langle 1 \mid \right) \left( \mid 1 \rangle \mid 2 \rangle \right) - \left( \langle 2 \mid \langle 1 \mid \right) \left( \mid 2 \rangle \mid 1 \rangle \right) \right),$$

$$= \frac{1}{2} \left( \langle 1 \mid 1 \rangle \langle 2 \mid 2 \rangle + \langle 1 \mid 2 \rangle \langle 2 \mid 1 \rangle \right.$$

$$\left. - \langle 2 \mid 1 \rangle \langle 1 \mid 2 \rangle - \langle 2 \mid 2 \rangle \langle 1 \mid 1 \rangle \right),$$

$$= \frac{1}{2} \left( 1 + 0 - 0 - 1 \right) = 0 \tag{5.10}$$

5. (a) Argue that $H$ is self-adjoint and $\mathbb{I}$ is self-adjoint, and then appeal to the fact (from Chapter 5) that the tensor product of two self-adjoint operators is self-adjoint.

Clearly, $\mathbb{I}$ is self-adjoint. In this basis given in this problem the matrix representation of $Z$ is given by:

$$Z = \begin{pmatrix} 1 & 0 \\ 0 & -1 \end{pmatrix},$$

and this matrix is clearly self-adjoint since $Z = Z^\dagger$.

(b) The eigenvalues and eigenvectors of the tensor product of two operators are given by the products of the eigenvalues of each operator and the tensor product of the eigenvectors of each operator. These results were covered in class and in the homework assignments. Therefore the eigenvalues of $Z \otimes \mathbb{I}$ are given by:

$$+1, +1, -1, -1.$$

The eigenvectors are given by:

$$|0\rangle |0\rangle, |0\rangle |1\rangle, |1\rangle |0\rangle, |1\rangle |1\rangle,$$

as can be seen from the following calculation:

$$Z \otimes \mathbb{I} \, |0\rangle |0\rangle = |0\rangle |0\rangle,$$

$$Z \otimes \mathbb{I} \, |0\rangle |1\rangle = |0\rangle |1\rangle,$$

$$Z \otimes \mathbb{I} \, |1\rangle |0\rangle = -|1\rangle |0\rangle,$$

$$Z \otimes \mathbb{I} \, |1\rangle |1\rangle = -|1\rangle |1\rangle.$$

(c) The spectral decomposition of $Z \otimes \mathbb{I}$ is given by:

$$Z \otimes \mathbb{I} = \underbrace{|0\rangle\langle 0| \otimes \mathbb{I}}_{P_0} - \underbrace{|1\rangle\langle 1| \otimes \mathbb{I}}_{P_1},$$

where $P_0$ is the projection onto the eigenspace corresponding to eigenvalues of $+1$ and $P_1$ is the projection onto the eigenspace corresponding to eigenvalues of $-1$.

(d) Consider the state:

$$|\phi\rangle = \frac{|0\rangle |0\rangle + |1\rangle |1\rangle}{\sqrt{2}} \tag{5.11}$$

The state will be entangled if we *cannot* find states $|\phi_A\rangle$ and $|\phi_B\rangle$ such that $|\phi\rangle = |\phi_A\rangle \otimes |\phi_B\rangle$. Letting

$$|\phi_A\rangle = a_0 |0\rangle + a_1 |1\rangle, \qquad |\phi_B\rangle = b_0 |0\rangle + b_1 |1\rangle$$

Then we have:

$$| \phi_A \rangle \otimes | \phi_B \rangle = (a_0 \, | \, 0 \rangle + a_1 \, | \, 1 \rangle) \otimes (b_0 \, | \, 0 \rangle + b_1 \, | \, 1 \rangle),$$

$$= a_0 b_0 \, | \, 0 \rangle \, | \, 0 \rangle + a_0 b_1 \, | \, 0 \rangle \, | \, 1 \rangle$$

$$+ \, a_1 b_0 \, | \, 1 \rangle \, | \, 0 \rangle + a_1 b_1 \, | \, 1 \rangle \, | \, 1 \rangle. \tag{5.12}$$

Then if $(5.11) = (5.12)$ we must have:

$$\frac{1}{\sqrt{2}} = a_0 b_0$$

$$0 = a_0 b_1$$

$$0 = a_1 b_0$$

$$\frac{1}{\sqrt{2}} = a_1 b_1. \tag{5.13}$$

If we examine the second equality in this list, $0 = a_0 b_1$ implies either $a_0 = 0$ or $b_1 = 0$. However, either of these conditions being satisfied is inconsistent with the first and the fourth inequalities in this list. So we cannot find states $| \, \phi_A \rangle$ and $| \, \phi_B \rangle$ such that $| \, \phi \rangle = | \, \phi_A \rangle \otimes | \, \phi_B \rangle$.

(e) Using the postulate of quantum mechanics concerned with measurement, if the outcome of Alice's measurement is $+1$ then the state collapses to:

$$\frac{P_0 \, | \, \psi \rangle}{\| \, P_0 \, | \, \psi \rangle \, \|} = \frac{(| \, 0 \rangle \langle 0 \, | \otimes \mathbb{I}) \left( \frac{1}{\sqrt{2}} (| \, 0 \rangle \, | \, 0 \rangle + | \, 1 \rangle \, | \, 1 \rangle) \right)}{\| \, (| \, 0 \rangle \langle 0 \, | \otimes \mathbb{I}) \left( \frac{1}{\sqrt{2}} (| \, 0 \rangle \, | \, 0 \rangle + | \, 1 \rangle \, | \, 1 \rangle) \right) \, \|}$$

$$= | \, 0 \rangle \, | \, 0 \rangle. \tag{5.14}$$

(f) Now if Alice obtained $+1$ for her measurement of $Z$ then the state collapses to $| \, 0 \rangle \, | \, 0 \rangle$. The measurement operator for Bob is:

$$\mathbb{I} \otimes Z = \underbrace{\mathbb{I} \otimes | \, 0 \rangle \langle 0 \, |}_{P_0'} - \underbrace{\mathbb{I} \otimes | \, 1 \rangle \langle 1 \, |}_{P_1'},$$

where $P_0'$ is the projection on to the eigenspace corresponding to eigenvalues $+1$ and $P_1'$ is the projection on to the eigenspace corresponding to eigenvalues $-1$. The probabilities that Bob measures $1$ or $-1$ are given by:

$$p \, (\text{Bob obtains} + 1) = \langle 0 \, | \, \langle 0 \, | \, P_0' \, | \, 0 \rangle \, | \, 0 \rangle = 1,$$

$$p \, (\text{Bob obtains} - 1) = \langle 0 \, | \, \langle 0 \, | \, P_1' \, | \, 0 \rangle \, | \, 0 \rangle = 0. \tag{5.15}$$

# Bibliography


M. Amaku, F. A. B. Coutinho, and F. M. Toyama. The normalization of wave functions of the continuous spectrum. *Revista Brasileira de Ensino de Física*, 42, 2020.

John S Bell. On the Einstein Podolsky Rosen paradox. *Physics Physique Fizika*, 1(3):195, 1964.

M. Belloni and R. W. Robinett. The infinite well and Dirac delta function potentials as pedagogical, mathematical and physical models in quantum mechanics. *Physics Reports*, 540(2):25–122, 2014.

D. Bohm. *Quantum theory*. Courier Corporation, 1951.

Max Born. Quantenmechanik der stoßvorgänge. *Zeitschrift für physik*, 38 (11–12):803–827, 1926.

Max Born, Hedwig Born, Irene Born, and Albert Einstein. *The Born-Einstein letters: correspondence between Albert Einstein and Max and Hedwig Born from 1916 to 1955*. Walker, 1971.

John F Clauser, Michael A Horne, Abner Shimony, and Richard A Holt. Proposed experiment to test local hidden-variable theories. *Physical review letters*, 23(15):880, 1969.

C. Cohen-Tannoudji, B. Diu, and F. Laloe. *Quantum Mechanics*. Wiley-VCH, 1992.

P. A. M. Dirac. A new notation for quantum mechanics. In *Mathematical Proceedings of the Cambridge Philosophical Society*, volume 35, pages 416–418. Cambridge University Press, 1939.

P. A. M. Dirac. *The principles of quantum mechanics*. Number 27. Oxford university press, 1981.

Jonathan P Dowling and Gerard J Milburn. Quantum technology: the second quantum revolution. *Philosophical Transactions of the Royal Society of London. Series A: Mathematical, Physical and Engineering Sciences*, 361(1809):1655–1674, 2003.

B. Friedrich and D. Herschbach. Stern and Gerlach: How a bad cigar helped reorient atomic physics. *Physics Today*, 56(12):53–59, 2003.

A. Furuta. One thing is certain: Heisenberg's uncertainty principle is not dead. *Scientific American*, 2012.

Daniel Garisto. The universe is not locally real, and the physics Nobel Prize winners proved it. *Scientific American*, 6, 2022.

François Gieres. Mathematical surprises and Dirac's formalism in quantum mechanics. *Reports on Progress in Physics*, 63(12):1893, 2000.

D. J. Griffiths and D. F. Schroeter. *Introduction to quantum mechanics.* Cambridge University Press, 2018.

Vincent P Gutschick and Michael Martin Nieto. Coherent states for general potentials. v. time evolution. *Physical Review D*, 22(2):403, 1980.

K. Hannabuss. *An introduction to quantum theory*, volume 1. Clarendon Press, 1997.

W. Heisenberg. *The physical principles of the quantum theory.* Courier Corporation, 1949.

W. Heisenberg. Über den anschaulichen inhalt der quantentheoretischen kinematik und mechanik. In *Original Scientific Papers Wissenschaftliche Originalarbeiten*, pages 478–504. Springer, 1985.

E. J. Heller. Guided Gaussian wave packets. *Accounts of chemical research*, 39(2):127–134, 2006.

Eric J. Heller. Time-dependent approach to semiclassical dynamics. *The Journal of Chemical Physics*, 62(4):1544–1555, 1975.

Eric J. Heller. Time dependent variational approach to semiclassical dynamics. *The Journal of Chemical Physics*, 64(1):63–73, 1976.

A. J. G. Hey, T. Hey and P. Walters. *The new quantum universe.* Cambridge University Press, 2003.

Daniel Huber, Eric J. Heller and Robert G. Littlejohn. Generalized Gaussian wave packet dynamics, schrödinger equation, and stationary phase approximation. *The Journal of chemical physics*, 89(4):2003–2014, 1988.

Lars Jaeger. The second quantum revolution. *From Entanglement to Quantum Computing and Other Super-Technologies. Copernicus*, 2018.

David M. Kaplan and F. Y. Wu. On the eigenvalues of orbital angular momentum. *Chinese Journal of Physics*, 9(1):31–33, 1971.

E. H. Kennard. Zur quantenmechanik einfacher bewegungstypen. *Zeitschrift für Physik*, 44(4-5):326–352, 1927.

John R Klauder and Bo-Sture Skagerstam. *Coherent states: applications in physics and mathematical physics*. World scientific, 1985.

Luca Magri, Peter J Schmid, and Jonas P Moeck. Linear flow analysis inspired by mathematical methods from quantum mechanics. *Annual Review of Fluid Mechanics*, 55(1):541–574, 2023.

B. Mielnik and O. Rosas-Ortiz. Factorization: little or great algorithm? *Journal of Physics A: Mathematical and General*, 37(43):10007, 2004.

Nevill Francis Mott. *An Outline of Wave Mechanics*. The University Press, 1930.

Dwight E Neuenschwander. *Emmy Noether's wonderful theorem*. JHU Press, 2017.

Michael Martin Nieto. Coherent states for general potentials. IV. three-dimensional systems. *Physical Review D*, 22(2):391, 1980.

Michael Martin Nieto and L. M. Simmons Jr. Coherent states for general potentials. III. nonconfining one-dimensional examples. *Physical Review D*, 20(6):1342, 1979a.

Michael Martin Nieto and L. M. Simmons Jr. Coherent states for general potentials. I. formalism. *Physical Review D*, 20(6):1321, 1979b.

Michael Martin Nieto and L. M. Simmons Jr. Coherent states for general potentials. II. confining one-dimensional examples. *Physical Review D*, 20(6):1332, 1979c.

H. P. Robertson. The uncertainty principle. *Physical Review*, 34(1):163, 1929.

E. Benitez Rodriguez, L. M. Arevalo Aguilar, and E. Piceno Martinez. A full quantum analysis of the Stern–Gerlach experiment using the evolution operator method: Analyzing current issues in teaching quantum mechanics. *European Journal of Physics*, 38(2):025403, 2017.

Jun John Sakurai and Jim Napolitano. *Modern quantum mechanics*. Cambridge University Press, 2020.

E. Schrödinger. An undulatory theory of the mechanics of atoms and molecules. *Physical review*, 28(6):1049, 1926.

B. Schumacher and M. Westmoreland. *Quantum processes systems, and information*. Cambridge University Press, 2010.

A. D. Stone. *Einstein and the quantum: The quest of the valiant Swabian*. Princeton University Press, 2015.

D. J. Tannor. *Introduction to Quantum Mechanics: A Time-dependent Perspective*. University Science Books, 2007.

Sin-Itiro Tomonaga and Takeshi Oka. *The Story of Spin*. University of Chicago Press Chicago, IL, 1997.

H. Weyl. Gruppentheorie und quantenmechanik, hirzel, leipzig. *Theory of Groups and Quantum Mechanics, 2nd ed.(1931), transl. H. P. Robertson, Dover, NY (1950)*, pages 100–101, 1928.

John Archibald Wheeler and Wojciech Hubert Zurek. *Quantum Theory and Measurement*, volume 40. Princeton University Press, 2014.

Wei-Min Zhang and Robert Gilmore. Coherent states: Theory and some applications. *Reviews of Modern Physics*, 62(4):867, 1990.

# Index